ASCE Manuals and Reports on Engineering Practice No. 83

Environmental Site Investigation Guidance Manual

Prepared by the Task Committee on
Hazardous Waste Site Assessment Manual of the
Environmental Engineering Division of the
American Society of Civil Engineers

D1500102

Published by the
American Society of Civil Engineers
345 East 47th Street
New York, New York 10017-2398

ABSTRACT:

This manual, Environmental Site Investigation, was prepared as a guidance on procedures for investigating and characterizing a site that may be or is contaminated with hazardous materials. It begins with the general overview of the evolution of the site assessment process from the original environmental audit through the passage of the Federal Superfund and related state Superfund legislation. The second chapter presents an extensive and detailed discussion on performing a Phase I site assessment. In Chapter 3 the Phase II intrusive site investigation process is discussed. Risk assessment issues are also included. The last two chapters present an overview of the Phase III (remedial investigation) and Phase IV (remedial planning, design, and implementation) procedures. While this manual's main purpose is to provide guidance to environmental consultants, it will also be used by clients, environmental lawyers, bankers, real estate managers, and others to evaluate the work completed by these consultants.

Library of Congress Cataloging-in-Publication Data

Environmental site investigation guidance manual / prepared by the Task
    Committee on Hazardous Waste Site Assessment Manual of the
    Environmental Division of the American Society of Civil Engineers.
        p. cm. — (ASCE manuals and reports on engineering practice : no. 83)
    Includes bibliographical references.
    ISBN 0-7844-0096-2
        1. Hazardous waste sites— Evaluation. 2. Hazardous wastes-Risk
    assessment. I. American Society of Civil Engineers. Task Committee on
    Hazardous Waste Site Assessment Manual. II. Series.
    TD1052.E58 1995   95-51216
    363.73'84'0973—dc20   CIP

MANUALS AND REPORTS ON ENGINEERING PRACTICE

(As developed by the ASCE Technical Procedures Committee, July 1930, and revised March 1935, February 1962, April 1982)

A manual or report in this series consists of an orderly presentation of facts on a particular subject, supplemented by an analysis of limitations and applications of these facts. It contains information useful to the average engineer in his everyday work, rather than the findings that may be useful only occasionally or rarely. It is not in any sense a "standard," however; nor is it so elementary or so conclusive as to provide a "rule of thumb" for nonengineers.

Furthermore, material in this series, in distinction from a paper (which expresses only one person's observations or opinions), is the work of a committee or group selected to assemble and express information on a specific topic. As often as practicable the committee is under the direction of one or more of the Technical Divisions and Councils, and the product evolved has been subjected to review by the Executive Committee of the Division or Council. As a step in the process of this review, proposed manuscripts are often brought before the members of the Technical Divisions and Councils for comment, which may serve as the basis for improvement. When published, each work shows the names of the committees by which it was compiled and indicates clearly the several processes through which it has passed in review, in order that its merit may be definitely understood.

In February 1962 (and revised in April, 1982) the Board of Direction voted to establish:

A series entitled 'Manuals and Reports on Engineering Practice,' to include the Manuals published and authorized to date, future Manuals of Professional Practice, and Reports on Engineering Practice. All such Manual or Report material of the Society would have been refereed in a manner approved by the Board Committee on Publications and would be bound, with applicable discussion, in books similar to past Manuals. Numbering would be consecutive and would be a continuation of present Manual numbers. In some cases of reports of joint committees, bypassing of Journal publications may be authorized.

MANUALS AND REPORTS OF ENGINEERING PRACTICE

#	
10	Technical Procedures for City Surveys
13	Filtering Materials for Sewage Treatment Plants
14	Accommodation of Utility Plant Within the Rights-of-Way of Urban Streets and Highways
31	Design of Cylindrical Concrete Shell Roofs
33	Cost Control and Accounting for Civil Engineers
34	Definitions of Surveying and Associated Terms
35	A List of Translations of Foreign Literature on Hydraulics
36	Wastewater Treatment Plant Design
37	Design and Construction of Sanitary and Storm Sewers
40	Ground Water Management
41	Plastic Design in Steel-A Guide and Commentary
42	Design of Structures to Resist Nuclear Weapons Effects
45	Consulting Engineering-A Guide for the Engagement of Engineering Services
46	Report on Pipeline Location
47	Selected Abstracts on Structural Applications of Plastics
49	Urban Planning Guide
50	Planning and Design Guidelines for Small Craft Harbors
51	Survey of Current Structural Research
52	Guide for the Design of Steel Transmission Towers
53	Criteria for Maintenance of Multilane Highways
54	Sedimentation Engineering
55	Guide to Employment Conditions for Civil Engineers
57	Management, Operation and Maintenance of Irrigation and Drainage Systems
58	Structural Analysis and Design of Nuclear Plant Facilities
59	Computer Pricing Practices
60	Gravity Sanitary Sewer Design and Construction
62	Existing Sewer Evaluation and Rehabilitation
63	Structural Plastics Design Manual
64	Manual on Engineering Surveying
65	Construction Cost Control
66	Structural Plastics Selection Manual
67	Wind Tunnel Model Studies of Buildings and Structures
68	Aeration-A Wastewater Treatment Process
69	Sulfide in Wastewater Collection and Treatment Systems
70	Evapotranspiration and Irrigation Water Requirements
71	Agricultural Salinity Assessment and Management
72	Design of Steel Transmission Structures
73	Quality in the Constructed Project-a Guide for Owners, Designers, and Constructors
74	Guidelines for Electrical Transmission Line Structural Loading
75	Right-of-Way Surveying
76	Design of Municipal Wastewater Treatment Plants
77	Design and Construction of Urban Stormwater Management Systems
78	Structural Fire Protection
79	Steel Penstocks
80	Ship Channel Design
81	Guidelines for Cloud Seeding to Augment Precipitation
82	Odor Control in Wastewater Treatment Plants
83	Environmental Site Investigation
84	Mechanical Connections in Wood Structures

COMMITTEE'S PURPOSE AND OFFICERS
Environmental Engineering Division
Hazardous Waste Site Assessment
Manual Task Committee

This manual was written by the Hazardous Waste Site Assessment Manual Task Committee of the Environmental Engineering Division of the American Society of Civil Engineers. The Task Committee's purpose and officers are presented herein.

Purpose: to write a manual describing the appropriate procedures for investigating and characterizing a site that may be or is contaminated with hazardous materials.

Office	Member	Affiliation
CHAIR:	Yee K. Cho, P.E., L.S.P.	CDW Consultants, Inc. 111 Speen Street, Suite 119 Framingham, Massachusetts 01701
VICE CHAIR:	Richard G. DiNitto, L.S.P.	ABB Environmental Services, Inc. Wakefield, Massachusetts
PAST VICE CHAIR:	Myron Rosenberg, PhD, P.E.	Camp Dresser & McKee, Inc. Cambridge, Massachusetts
SECRETARY:	Joseph S. Devinny, PhD	University of Southern California Los Angeles, California

MAJOR CONTRIBUTORS

The following individuals are major contributors to the preparation and final production of this manual. The lead authors prepared the preliminary and final drafts of their respectively assigned chapters as recognized experts in their disciplines. The members of an appointed Peer Review Subcommittee provided their comments on the preliminary draft of the document, which comments were incorporated into the final draft by the lead authors. The appointed ASCE Blue Ribbon Committee, which membership represented leading national experts throughout the country on oil and hazardous material contaminated sites issues, offered their critique of the final draft of the manual. The Technical Editors evaluated the comments received from the Blue Ribbon Committee and incorporated appropriate comments into this final edition.

Lead Authors:

Introduction:	Yee Cho, P.E., L.S.P.	CDW Consultants, Inc., Massachusetts
Chapter 1:	Daniel Buss, P.E.	Camp Dresser & McKee, Inc., Wisconsin
Chapter 2:	Dale Stirling	Landau Associates, Inc., Washington
Chapter 3: (final)	Carol Whitlock, P.E.	B & V Waste Science and Technology Corp., Missouri
Chapter 3: (preliminary)	Richard DiNitto, L.S.P.	ABB Environmental Services, Inc., Massachusetts
Chapters 4 & 5:	Joseph Devinny, Ph.D.	University of Southern California, California

Blue Ribbon Committee:

Frederick Boecher	U. S. Army, Maryland
Terence McManus	Intel Corporation, Arizona
N. Shashidhara, Ph.D.	Raytheon Corporation, New Jersey
Udai Singh	CH2M Hill, California

Robert Williams, P.E.   Agency for Toxic Substances and Disease Registry,
                         Georgia
Doreen Zankowsky,   Camp Dresser & McKee, Inc., Massachusetts
Esq.

Technical Editors:

All Chapters: Yee Cho, P.E., L.S.P.   CDW Consultants, Inc.,
                                                Massachusetts
Chapters 1 & 2:   Richard DiNitto, L.S.P.   ABB Environmental
                                                Services, Inc.,
                                                Massachusetts
Chapters 3, 4, & 5: Carol Whitlock, P.E.   B & V Waste Science and
                                                Technology Corp.,
                                                Missouri

ACKNOWLEDGEMENT

This manual has been prepared under the auspices of the ASCE Environmental Engineering Division and under the leadership of the following Division chairs. Their total support was crucial to the successful completion of this manual.

Environmental Engineering Division Chairs:

Dr. Terence McManus
Dr. Michael Saunders
Dr. David Stensel
Dr. Cecil Lue-Hing

Also, the following individuals contributed valuable comments to the preliminary and/or final draft documents:

Kathleen Campbell	CDW Consultants, Inc., Framingham, Massachusetts
Wendy Cohen	Davis, California
Kerry Converse	CDW Consultants, Inc., Framingham, Massachusetts
Julius Diogenes, P.E.	CDW Consultants, Inc., Framingham, Massachusetts
Farooq Siddique, P.E.	CDW Consultants, Inc., Framingham, Massachusetts
John Walker, P.E.	CDW Consultants, Inc., Framingham, Massachusetts

In addition, the immense support demonstrated by all in the general community, the environmental engineering community, the site investigation disciplines, and the Committee members and officers was the driving force toward the completion of this manual. Finally, the commitment by the Chapter Authors, the Blue Ribbon Committee members, the Technical Editors, and the respective affiliations of each of these individuals were essential toward the fruition of this manual.

To all those who have contributed to this manual via written word or in spirit, this manual is hereby dedicated.

(Ms.) Yee K. Cho, P.E., L.S.P.
Chair
Hazardous Waste Site Assessment Manual Task Committee

March 31, 1995

Contents

INTRODUCTION

This manual has been prepared as a guidance document on procedures for investigating and characterizing a site that may be or is contaminated with hazardous materials. As written, the targeted audience is primarily environmental consultants, although it can be used and is expected to be used by clients, environmental lawyers, bankers, real estate managers, financial institutions, the judicial system, and others to evaluate the work products completed by environmental consultants.

In Chapter 1, this manual begins with the general overview of the evolution of the site assessment process from the original environmental audit through the passage of the Federal Superfund and related state Superfund legislation. Chapter 1 also reviews and compares a very limited number of site assessment manuals developed by other organizations. It details the major difference between this ASCE manual and those of other organizations: the ASCE manual delineates the Phase II subsurface investigation processes; the other manuals end with the Phase I preliminary assessment process. The chapter further provides a general overview of the upcoming chapters in the document. Furthermore, the chapter presents a major discussion on the legal issues that consultants involved in site investigation need to be familiar with and be alerted to; it describes contracting vehicles as well as the language recommended for contracts executed by the consultants with their clients. Finally, a sample Phase I report format is included in this chapter.

Chapter 2 presents an extensive and detailed discussion on performing a Phase I site assessment. It also details liability issues for potentially responsible parties and defenses available to innocent landowners. Some of the procedures described for the Phase I assessment are compared with those published by others, such as the American Society for Testing of Materials (ASTM).

Chapter 3 is much more detailed and longer than the other chapters because the Phase II intrusive site investigation process is so much more extensive with a wider diversity of options than the Phase I process discussed in Chapter 2. The procedures discussed in Chapter 3 are not included in the above ASTM (and others) document on site assessment and, therefore, no similar type of comparison, as conducted in Chapter 2, is presented. Chapter 3 is a relatively technical section because of the nature of engineering and scientific processes. Risk assessment issues are also discussed in this chapter. Although this chapter presents a very detailed summation of the available options and the procedures to conduct a Phase II investigation, it is still not and cannot be all inclusive. The engineering, science, and art of the Phase II investigation is still evolving, with constantly changing technologies and regulations governing the same. A sample Phase II report format is also included in this chapter.

Chapter 4 and Chapter 5 were prepared as only cursory chapters. These chapters do not contain the full substance of the Phase III and Phase IV processes as written for Phase I and Phase II in Chapter 2 and Chapter 3, respectively. These final two chapters are only intended to provide a complete picture of the site investigation process and to present only an overview of the Phase III and Phase IV procedures in this manual. These final two chapters will be presented in full detail in Volume II to this manual which has an anticipated publication date of 1997.

All chapters of this document have been presented and published at national and international conferences and conventions when the draft versions of the document were prepared. A draft of the *Manual* has also been presented by the officers and lead chapter authors at various local meetings and seminars throughout the country. A listing of some of the published articles are included in the reference.

Also included in the references for this introductory section are the Massachusetts State regulations and its policies and guidelines relative to site investigation for oil and hazardous material contamination. Numerous reference documents published by the U.S. Environmental Protection Agency are listed in each chapter of this manual. The Massachusetts regulations and policies are included since they are very detailed in their descriptive processes of what needs to be done to conduct a site investigation according to the requirements of the state environmental agency. Also, they are listed herein since the site investigation program in Massachusetts is relatively progressive in terms of having established and detailed mandatory procedures in place. Finally, Massachusetts was the first state to establish a licensing process for professionals involved in conducting environmental site investigation (i.e., Licensed Site Professional program) in accordance with the state's regulations.

In closing, this document is not intended to be all inclusive. It is also not intended to be a document of standards of the environmental site investigation processes, especially since this environmental discipline and its governing regulatory requirements are dynamic and still evolving. Other editions of this manual are planned to incorporate changes in this discipline through the years and to include appropriate comments from the readers. Comments on this document can be directed to the current Chair of the Hazardous Waste Engineering Committee of the ASCE Environmental Engineering Division.

REFERENCES

Buss, D. and Cho, Y., "ASCE Environmental Site Investigation Compared," *CSCE-ASCE International Environmental Engineering Conference Proceedings*, July 1993.

Buss, D., Rosenberg, M., Barker, F., Capriotti-Hesketh, A.M., Ma, Y.Y., and Cho, Y., "The Relationship Between Environmental Site Investigation Protocols of ASCE, ASTM and AGWSE Guidance Documents," *Proceedings of HMC Superfund '92 National Conference*, December 1992.

Cho, Y., "ASCE Draft Environmental Site Investigation Guidelines," abstract for National Ground Water Association Environmental Site Assessment Conference, 1992.

Cho, Y. and Stirling, D., "Critical Importance of History in Phase I Environmental Site Assessments," *Proceedings of HMC/Superfund '92 National Conference*, December 1992.

Devinny, J. and Lu, J.C.S., "Environmental Site Investigation Manual: Remedial Investigation and Remedial Planning, Design, and Implementation," *CSCE-ASCE International Environmental Engineering Conference Proceedings*, July 1993.

DiNitto, R., Varoutsos, B.M., and Cho, Y., "What Everyone Needs for Completion of a Phase II Site Investigation in an Environmentally Sound and Cost-Effective Manner," *Proceedings of HMC/Superfund '92 National Conference*, December 1992.

Massachusetts Contingency Plan, 310 Code of Massachusetts Regulations 40,000, January 1995.

Policies and guidelines of the Massachusetts Department of Environmental Protection Bureau of Waste Site Cleanup, various dates.

Oil and Hazardous Materials Site Assessment, Standards Procedures Manual, CDW Consultants, Inc., Framingham, MA, 1992.

Chapter 1

BACKGROUND AND LEGAL ASPECTS

1.1 INTRODUCTION

The purpose of Chapter 1 is to present 1) a brief history and origin of the environmental investigation process regarding real property transactions, 2) a general description of environmental assessment standards developed by other technical societies, 3) the ASCE's phased environmental site investigation process, and, 4) an overview of important legal and risk management concerns between the consultant and client regarding environmental site investigations.

The level of perception and sophistication of the environmental investigation process increases from day to day while many of the procedures for performing these inherent tasks are still evolving. This sophistication is not only driven by the development of new procedures to address a greater understanding of environmental relationships in the technical community but also by legal, financial, and political requirements involving real property transactions, as well as by the need to evaluate the extent of contamination due to accidental releases, and the need to meet state and federal requirements. One of these issues, the legal relationship between the consultant and client, will be discussed since responsibilities of both of the aforementioned entities must be fulfilled to achieve the goals of the environmental investigation process as described in this manual.

1.2 TECHNICAL AND PHILOSOPHICAL APPROACH

Several organizations and agencies, including the American Society for Testing of Materials (ASTM), the Association of Ground Water Scientists and Engineers (AGWSE), and the U.S. Environmental Protection Agency (EPA), have formulated technical and procedural approaches regarding the performance of environmental site assessments. In general, most approaches have been directed toward a limited environmental assessment (commonly referred to as a Phase I Site Assessment, or File Review and Site Walkover) to satisfy potential liability issues with the Comprehensive Environmental Response, Compensation, and Liability Act (CERCLA).

Environmental site investigations (ESIs) have become an integral part of the due diligence process when acquiring or selling real property, regardless of the size of the transaction. A systematic approach emphasizing thorough planning, project management, coordination, and standardization of all elements of work should be utilized when performing an ESI. Many problems develop through inadequate communication and insufficient attention to project management and coordination. Through a systematic approach to project management and coordination, many of the problems that could arise in meeting project commitments would not occur or would be quickly recognized allowing action to be taken before there is an impact in the ESI.

There are many skills that must be drawn upon to properly conduct the various activities of an ESI. As the ESI process (broken into four phases in this *Manual*) moves from the limited assessment through invasive subsurface investigation techniques, and in some cases to remedial design and cleanup, the shift in skills required will be from the scientific and engineering disciplines to the engineering and construction disciplines. The engineer must play an active role in all phases of work to provide for the gathering of applicable data for use in the selection, design, and implementation of corrective measures. This *Manual* will discuss the expertise required in the four phases of work and the means by which a team is selected to best address environmental conditions that are discovered. This expertise will be addressed in each chapter of this manual. The review of information and reports generated during the environmental assessment process by qualified staff is an integral part of the process providing for quality control and quality assurance at each and every step of the process.

1.3 HISTORY OF ENVIRONMENTAL ASSESSMENTS

During the early 1970s, numerous environmental regulations were promulgated. As these regulations grew in complexity, a process called environmental auditing evolved to serve as a tool for managing environmental compliance with the regulations. In general, the environmental auditing process laid the groundwork for the environmental assessment and investigation process as it is known today. These processes evolved in response to the various environmental laws that were developed and the resultant liabilities associated with the ownership of contaminated property.

The experience of companies with corporate environmental auditing programs has shown that a sound auditing program enables an organization to:

- identify and correct environmental problems, and reduce the potential for criminal and civil liability, litigation, and/or fines;
- assure accurate cost projections to meet U.S. Security and Exchange Commission (SEC) regulations that require publicly-held companies to disclose current and projected costs of compliance with environmental regulations;
- acquire critical information for financial planning (including re-financing and debt restructuring needs), mergers, and potential corporate expansions; and
- organize compliance information in ways which both promote compliance and effectively capitalize on related opportunities (e.g., recycle, reuse, conservation of raw materials, or emissions trading).

The environmental audit gained acceptance as corporations recognized its attributes as a management tool.

In the latter 1970s and 1980s, EPA and several state agencies examined the environmental auditing approach relative to their regulatory programs. The EPA's Office of Policy and Resource Management at one time reviewed a trial implementation of the environmental audit through a limited number of state pilot programs. The pilot programs would be implemented in those responsible states expressing interest. EPA would assist the states in developing an acceptable program that satisfied state and federal requirements and served both industry and the general public. Facilities with reliable compliance records and sound environmental auditing procedures could volunteer to participate. Approved facilities would be encouraged to establish, maintain, or upgrade internal audit procedures.

As an example, in Michigan an advisory task force comprised of members representing industry, local governments, environmental groups, and citizens groups was assembled to evaluate the environmental auditing concept. (Tanner, 1982). Most support was voiced for a voluntary environmental auditing program, with the Michigan Department of Natural Resources (MDNR) providing technical assistance to companies interested in establishing this type of program. (MDNR, 1983).

The conceptual approach within EPA has changed significantly between the mid-1970s and the present. (Blumenfeld, 1982). The earliest auditing concepts called for certified third party auditors. Firms would retain auditors that would certify to the regulatory agencies that permits and regulatory requirements were adhered to. This concept was ultimately rejected by EPA.

The second approach was incentive based. This approach was built on private-sector auditing already in place. The rationale consisted of five major elements.

- In-house auditing was economically positive and socially responsible.
- Because an in-house program scrutinizes its facilities more closely than an inspector, more noncompliance problems could potentially be identified.

- Firms with in-house auditing programs were more likely to be in compliance.
- Corporate management had a strong interest in encouraging environmental staff to know and understand potential environmental problems and act on these problems.
- EPA risks related to non-compliance problems would be minimized.

The present three-pronged approach taken by EPA is to 1) endorse private-sector auditing efforts, 2) get a better understanding of environmental auditing, and 3) provide substantial assistance to interested states and firms. EPA believes this approach should be more acceptable to industries because it does not involve the potential regulatory apparatus of eligibility criteria, standards of adequacy, and formal certification. In addition, it will allow industries to tailor their programs to their specific management style and organizational structure.

While environmental auditing provides a valuable tool to assist firms in preventing potential environmental impacts from current operations, it does not address prior releases to the environment, or contamination on abandoned properties. Nor does auditing provide for a systematic or technical approach to evaluating a problem once it is created. Combined with an increased awareness of the liability associated with owning or operating a contaminated property, the public, the regulators, and the regulated community turned to new legislation, regulation, and a process for assessing the potential for liability associated with a property.

1.3.1 Environmental Assessments

Since the enactment of the CERCLA in 1980, previous and current owners and operators of a property can be held strictly liable for contamination on their property regardless of how long the situation existed or who was responsible for creating the contamination. The Superfund Amendments and Reauthorization Act of 1986 (SARA) offers a provision for a landowner to establish its innocence regarding knowledge that on-site contamination had occurred. To establish this defense, the owner must demonstrate that, at the time of the acquisition, the owner did not know or have reason to know that hazardous substances were released or disposed of on the property. To preserve that defense, the owner must demonstrate that he carried out "all appropriate inquiry into the previous ownership and uses of the property consistent with good commercial and customary practice in an effort to minimize liability." (CERCLA, §101(35)(b)). The need for this language and similar protection continue to be driven by the financial lending community, as they typically hold the mortgage on the property and may be held responsible. However, it must be noted that the lending institution does not typically engage in or control the day-to-day management of the business that operates on the property. The distinction between the owner and the operator and the mortgagor cannot be understated and must be clearly understood.

In response to liability under the CERCLA legislation and to provide protection under the SARA provisions, various technical associations and societies have developed guidance or standards for conducting appropriate inquiry into the environmental status of property. Two of these societies and associations, in addition to the American Society of Civil Engineers (ASCE), are the American Society for Testing of Materials (ASTM) and the Association of Ground Water Scientists and Engineers (AGWSE). ASTM and AGWSE guidelines or standards are focused only on the Phase I process with no invasive subsurface investigation. This ASCE guideline details the process to conduct a Phase II subsurface investigation, and provides preliminary summaries on the Phase III remedial investigation and Phase IV remedial planning, design, and implementation.

1.3.1.1 ASTM Standards for Environmental Assessments   Briefly, the ASTM presents two standards for assessing property prior to an ownership transaction. ASTM Standard Practice E.50.02.2 defines the commercial and customary practice for conducting a Phase I environmental site assessment. ASTM E.50.02.1 presents the Standard for the Transaction Screen Process. The "Transaction Screen Process" may allow the user to conclude that no further inquiry is needed to assess the potential for ". . . identifying a Recognized Environmental Condition at the property, and have the performance of the Transaction Screen Process constitute appropriate inquiry without undertaking the Phase I environmental site assessment." (ASTM Standard E.50.02.1, 1992).

1.3.1.2 AGWSE Guidance for Environmental Assessments   Similar to the ASTM, the AGWSE limited their guidance to the performance of Phase I environmental site assessment tasks. This guidance is presented in the document titled *Guidance to Environmental Site Assessments*, September 1992. Unlike the ASTM, where Phase I tasks have been included as definitive standards, the AGWSE guidance document was developed to enable the investigator and client to choose tasks from a "menu of technical options" to tailor the investigation to best identify potential environmental concerns. (AGWSE, 1992).

1.3.2 ASCE Environmental Site Investigation Process

Whereas other organizations have limited their guidance to the initial or Phase I assessment, the ASCE environmental site investigation (ESI) process, as presented in this *Manual*, addresses activities involved in the multi-media evaluation of a designated land area for purposes of determining the presence or absence and distribution of potential contaminants that pose, or could pose, a threat to human health and the environment or that might preclude certain uses of the property. This process forms the basic structure of this *Manual*.

Some of the goals of this *Manual* are to develop a guidance document that includes references to existing documents and textbooks on related topics to use as ready references for the reader, to prepare a guidance manual that is user friendly, and to present procedures for the preliminary investigation phase to the implementation of corrective measures. This *Manual* includes a review of guidance documents prepared by other organizations, and incorporates in this manual, by reference, items that are appropriate for use in the manual.

This ESI *Manual* follows a phased approach that creates milestones by systematically and logically separating tasks by decision. These milestones allow for control of work activities and minimization of costs by providing decision points within the complete ESI process. The major phases of the ESI are:

- Phase I—Preliminary Site Assessment
- Phase II—Site Investigation
- Phase III—Remedial Investigation
- Phase IV—Remedial Planning, Design, and Implementation

Phase I is a nonintrusive data search and review combined with an on-site inspection. This phase of work involves the review of site-specific information, a site walkover, and interviews with people familiar with site operations and with agency personnel knowledgeable of the facility's compliance with environmental regulations. Chapter II discusses the tasks generally performed under the preliminary site assessment process (Phase I).

Phase II work generally involves invasive or intrusive testing coupled with laboratory analysis. The media tested will depend on site specific conditions identified in Phase I. Media to be considered for testing include, but are not limited to, soil, groundwater, surface water, air, and substances at the property that require identification for purposes of proper management. Chapter III presents Phase II site investigation task elements.

Phase III consists of the remedial investigation as defined by EPA and CERCLA guidelines. This phase is generally an extension of the Phase II tasks. This phase involves the collection of data defining the areal extent of contamination, its mobility, and potential risk to human health and the environment. Information to be used in the latter remedial planning and design phase is generated during Phase III. Chapter IV of this *Manual* preliminarily and generally discusses the remedial investigation process.

Phase IV involves remedial planning such as evaluating treatment alternatives based on their feasibility for implementation, development of design documents, and implementation of the corrective action plan. Phase IV also presents protocols for verifying cleanup activities and developing the site closure report. Chapter V of this *Manual* preliminarily and generally discusses remedial planning, design, implementation of corrective actions, and site closure activities.

1.4 LEGAL RISK MANAGEMENT ASPECTS OF THE ENVIRONMENTAL SITE INVESTIGATION PROCESS

Environmental investigations should follow a systematic approach consisting of planning, information and data gathering, analyzing and reviewing information, and reporting of the assessment findings. Certain project performance issues that could occur during all phases of the work can be addressed by recognizing these potential issues at the contractual stage of the project and incorporating appropriate language in the scope of work and contract documents. The following areas of the environmental investigation process account for most of the potential problem issues: (Buss, 1992).

1. establishing the contractual relationship;
2. assuring cooperation with property personnel and current owner(s)/operator(s);
3. adhering to work plan specifications;
4. obtaining sufficient information; and
5. maintaining good documentation of the sources of information and of observations made by the consultant's personnel.

Exposure to these potential problem areas can be minimized through proper planning, coordination, and management control of the environmental investigation. The following discussion addresses these areas and the management controls that can be implemented during the environmental investigation to prevent their occurrence or minimize their impacts.

1.4.1 Contractual Relationship

There are two basic risks imposed by environmental laws, cleanup liability for contaminated properties, and compliance liability for violating environmental laws. Developing an agreement that covers both the client's and the consultant's concerns related to these risks can sometimes be time consuming and difficult. It is essential, however, that Phase I through Phase IV services be performed under an agreement that specifies the responsibilities of all parties involved in an ESI. The following issues should be addressed to provide adequate understanding and protection to all parties involved in the ESI.

1.4.1.1 Site and Property Access Agreements   When contracting with an environmental consultant to perform an ESI, one major factor is granting the consultant access to the site and/or assisting the consultant with obtaining site access. Site access is a two-fold issue. It not only means physical access to the property but also to specific areas within the site. The following examples illustrate this issue.

- If the client is also defined as the owner of the site, then site access should be easily granted. The client/owner would allow the consultant access through a provision in the contract. Additionally, site access should be defined in the contract as limited to areas that the consultant can actually enter (e.g., an area behind a locked door cannot be considered accessible to the consultant).

• If the client is not the owner of the site, then the client must either obtain access for the consultant or assist the consultant in obtaining site access. Either way, if the client cannot obtain the access for the consultant, this should be considered a *force majeure*-type event and all legal relief offered under standard *force majeure* contractual language should be allowed. In other words, the consultant cannot complete the scope of work due to conditions beyond the consultant's control.

In both cases, the consultant must be allowed to qualify his report to the client by explicitly defining areas where access could not be obtained. The following is an example of access language that can be used for ESI related contracts.

> Owner shall arrange for access to and make all provisions for consultant to enter upon public and private property as required for consultant to perform its services under this Agreement.

1.4.1.2 *Confidentiality Agreements*   The only reliable mechanism to protect information generated by the consultant from disclosure is a confidentiality agreement. A confidentiality agreement can be a separately signed document or can be included as a clause in the consultant's agreement binding the consultant to protecting the confidentiality of the information disclosed to the consultant or generated during the project. A confidentiality agreement should explicitly exclude any and all information which may already be known to the public at the time of signing the agreement.

Additionally, even if the consultant is contracting with the purchaser or lender, the third party (the seller or borrower) may want to demand a confidentiality agreement with the purchaser (or lender) that includes the consultant as a signatory. In this way, the property owner makes the decision whether it is appropriate to notify regulatory agencies of the information discovered. Contractual protection is necessary because no common law duty or restriction prevents disclosure. Without a three-cornered agreement including the consultant, the purchaser, and the seller, there is no legal mechanism to prevent the consultant, purchaser, or lender from making information public.

Standard confidentiality language for a contract includes:

> Consultant shall keep all information and data furnished and obtained in connection with the project confidential except when such information is requested by a court of law or legal administrative process.

The language can be broadened by requiring the consultant to notify the owner immediately in the event of any such court order. This will give the owner ample time to file for a protective order, if warranted.

It is unlikely that an effective mechanism can be utilized to protect against the disclosure of assessment information in legal proceedings, because the attorney-client privilege and attorney work-product doctrine probably will not apply. If the purchaser or lender is the client, even the use of an attorney as the go-between does not necessarily protect the information. If the work does not concern any of the property of the client, it would be difficult to argue that the information is part of the legal advice rendered to the client. Similarly, the attorney work-product doctrine only protects information generated in the course of pending or threatened legal proceedings. It can hardly be argued that a purchase or loan transaction is such a proceeding.

Because protection of assessment information is an important issue in due diligence practice today, facility owners and operators can and have implemented the following:

- retain independent environmental consultants through company or outside legal counsel, as consulting experts to prepare for litigation concerning compliance matters,
- structure the assessment or audit as an investigation by company or outside legal counsel, with the assistance of environmental consultants,
- require the consultants to deliver the report only to company or outside legal counsel, with all of the consultant's notes and working papers,
- communicate the results orally in meetings between company managers and counsel, and
- undertake follow-up work and corrective action at the direction of counsel based upon recommendations from the consultant.

In all cases, the consultant cannot protect the information if it is in the consultant's possession. The consultant can, if requested by the client, notify the client immediately upon receiving a court or administrative order. The client can then file for a stay or protective order, or the client can fight the request on legal grounds.

1.4.1.3 Client's Responsibilities When contracting with a client, the consultant must have contractual protection that the client will perform certain activities and/or functions. The most important functions when conducting ESIs are for the client to provide all existing information to the consultant, designate a person to act on the client's behalf, sign any and all required manifests (in the unlikely event it should be necessary), promptly report regulated conditions to the appropriate public authorities in accordance with applicable law, assume responsibility for personal and property damage caused by interference with subterranean structures that are not shown on documents provided to the consultant, and bear all costs incident to the client's contractual responsibilities.

Sample contract language is offered below.

> Owner shall do the following in a timely manner so as not to delay the services of consultant:
>
> a. provide all criteria and full information as to Owner's requirements for the project and designate a person with authority to act on Owner's behalf,
> b. furnish to the consultant all existing studies, reports and other available data, and services of others pertinent to the project; authorize the consultant to use and rely upon such information,
> c. sign any and all required manifests relating to the generation, transportation, storage, treatment, and disposal of all wastes associated with the Agreement,
> d. promptly report regulated conditions, including without limitation, the discovery of releases of hazardous substances at the site to the appropriate public authorities in accordance with applicable law,
> e. agree to assume responsibility for personal and property damages due to Consultant's interference with subterranean structures such as pipes, tanks and utility lines that are not correctly shown on the documents and information provided by Owner to Consultant, and
> f. bear all costs incident to compliance with the requirements of this section of the Agreement.

1.4.1.4 *Consultant's Responsibilities* In addition to the client having contractual responsibilities, the consultant must be bound to perform certain duties as well. One of the most important responsibilities of the consultant is to perform its services in accordance with the "Standards of the Environmental Consulting Profession." The following clause should be included in the Agreement.

> The services provided by Consultant shall be performed in accordance with generally accepted professional engineering practice at the time when and the place where the Consultant's services are rendered. Consultant's services shall not be subject to any express or implied warranties whatsoever.

Additionally, the consultant shall provide its services in accordance with the laws of the state in which the project is located, protect the safety of its employees and the employees of its subcontractors, prepare a report of its findings for the client, and promptly notify the client of any issues requiring reporting to the appropriate public authorities.

1.4.1.5 Risk Sharing The basic concept underlying the recommended contract provisions for indemnity and limitation of liability contained herein is based on the fairness in allocating risk and benefit between the client and the consultant providing the services. Some of the factors evaluated in arriving at a fair position are:

- availability of risk-shifting mechanisms such as insurance,
- benefit (such as profit) for risk taken,
- control over the entire process or project, and
- capacity to absorb the risk.

Each project should be evaluated on its own merits and the recommended contractual language should reflect risks accordingly.

1.4.1.6 Indemnification Indemnification can be defined as the making whole of another person for injury or damage done to that person or piece of property. Indemnification clauses often contain obligations that go beyond indemnification protection as is commonly understood under common law. An agreement to "defend" the other party, for instance, may require the paying out of money to the other party's defense attorney at a much earlier point in time than if the clause had required one to "indemnify for the cost of defense."

An agreement to indemnify the client by the consultant creates serious obligations that do not exist in the absence of such a clause. Probably the most serious of these is the potential loss of the workers' compensation shield.

Factors involved in an ESI make it mandatory for the prudent consultant to aggressively seek indemnification from the client to the maximum degree allowed by law both for claims arising out of the consultant's services and for claims arising out of the work of others. The indemnification position sought is total indemnification of the consultant by the client except for the consultant's willful misconduct or gross negligence.

Many clients will not accept such indemnification language. The alternative should be that of mutual indemnity for the "negligent acts, errors, or omissions" of either party—in essence, a comparative indemnity. Additionally, the consultant should seek full indemnity from the client for any "releases or threatened releases of pollutants."

Although it would be best to receive full indemnity from the client, the alternative method stated above (comparative indemnity with a limitation of liability running to the consultant) would be the prudent approach, especially since indemnification for the professional consultant's **negligent performance** may be legally impermissible in some states which have "anti-indemnification" statutes. Legal counsel should be consulted concerning the applicable law in the state where the work will be performed as well as the state whose law governs the interpretation of the agreement. "Anti-indemnification" statutes must be reviewed together with the status of the two parties (e.g., public vs. private sector). It is unreasonable for a client to demand, and foolish for an environmental consultant to agree to, indemnification of the client for the consultant's non-negligent acts in an ESI-type project.

1.4.1.7 *Limitations of Liability* Another method of limiting the exposure of the consultant would be to seek a "Limitation of Liability" from the client. This is an important alternative due to the uninsurable and financial risks involved in the ESIs. The consultant can ask that "total liability under this Agreement be limited to the fees earned under the Agreement or a specified monetary amount." The specified amount would usually be developed through negotiations based on the risks involved with the project and is commonly set as some multiple of the contract value, not to exceed a specified upper limit. The amount of profit realized by the consultant under the ESI project is *de minimus* when compared to the liability exposure associated with this type of work.

1.4.1.8 *Schedule* It is best if the schedule in the agreement is mutually agreed to by the consultant and the client. The consultant may have to obtain regulatory reviews and/or rely on others for information and this could result in a delay in the consultant's performance (a delay beyond the consultant's control). Language to be avoided includes, "Time is of the essence in the performance of services under this Agreement." This connotes Uniform Commercial Code (UCC) provisions and could result in a charge of breach of contract if the schedule is not met.

Contractual language should provide for some flexibility in the schedule should the delay be beyond the consultant's control. A standard *force majeure* clause would take care of this issue and should be included in the Agreement. Additionally, language should be included that the "Consultant will be relieved of its schedule obligations and the schedule will be adjusted accordingly upon notification by Consultant to the Owner of the reasons for such adjustment."

1.4.2 Use of Information

The reason for including a disclaimer on an ESI report is to discourage its use up and down the transaction chain. The consultant is hired by one client and the consultant's contractual duties run solely to that client. If the client chooses to allow others to rely on the report, that is the client's sole decision (he bought and paid for the report).

Additionally, it is prudent to limit the "shelf-life" of the report. Site conditions change and if the report is used for anything other than its intended purpose and after a long delay (e.g., sitting on a shelf for more than 30 days), conditions could change. The consultant would not want to be legally liable for any changes that the consultant did not personally observe or study that occurred after the submittal of the report.

The ESI report should include a standard "Statement of Use" clause, such as the following, that covers these two items.

This report has been prepared by (*Insert Consultant's Name*) for the exclusive use of (*Insert Client's Name*) and the information, results, and conclusions contained in the report are valid for a period of thirty (30) days from the date of the report. This report is not intended or represented for use by any other party. Any use of this report beyond the thirty (30) day period or use by others at anytime is at the sole risk of the user.

In addition to disclaimers, the consultant should make sure that the conclusions in the report are "qualified." That is, "the environmental audit conducted by (*Insert Consultant's Name*) did not reveal any apparent, observable environmental contamination problems." Or, "the Consultant's ESI revealed. . . ."

1.4.3 Certifications

Clients and regulatory agencies are increasingly requesting or requiring that the consultant provide certifications that the project has been completed in accordance with the plans and specifications or a certification that the site is clean in the instances of an ESI. Providing such a certification exposes the consultant or the consulting firm to excessive liability.

General principals of law require that a professional such as an engineer or environmental consultant perform its services in accordance with the standards of the profession in a non-negligent manner. Professionals may be responsible for damages which arise out of their negligence. The law does not impose a standard of perfection on professionals. When a consultant provides a certification, however, he in effect imposes liability on himself that would not ordinarily exist. A second concern with certifications is the express exclusion contained in all professional liability insurance policies. Professional liability insurance policies explicitly state that no coverage will be provided for any warranty, guarantee, or certification provided by the consultant. This means that the assets of the firm may be at risk if the consultant provides a certification.

If the client adamantly insists or the state law requires a certification, the following approach and language could be used.

The word "certify" as used in this statement is understood to be the professional opinion of the consultant which is based upon the consultant's knowledge, information, and belief, formulated in accordance with commonly accepted procedures consistent with applicable standards of practice, and as such does not constitute a guarantee or warranty, either express or implied.

If the client will not accept the above approach for "certification," the consultant should contact its legal counsel for advice. In summary, understanding the risk and negotiating a contract that will protect the consultant, and yet be fair to the client, is a major undertaking. However, it need not be a difficult undertaking.

1.4.4 Cooperation and Communications

Prior to on-site work, the designated owner's representative at each property should be contacted by the client's project coordinator. This initial contact by the client is important to introduce the consultant personnel, present the purpose of the investigation, and ensure cooperation with all involved parties. Without a cooperative spirit, the investigation cannot proceed since access to the property and available information may not be readily obtained.

A standard list of questions should be presented by the consultant to each designated owner's representative at the property at the initiation of the project. Questions should be directed to the owner's representative to determine health and safety requirements, sensitivity issues, schedule, availability of information, and site accessibility. Following the aforementioned discussion, a letter should be sent to each property owner's representative confirming the consultant's request for information and the schedule for the on-site inspection. This letter should also be directed to the client's environmental project coordinator to ensure that all interested parties are aware of project requirements. The property owner's response to the request for information will enable the consultant to establish what information is available for review and the need for supplemental information.

1.4.5 Assessment Protocol

The on-site inspection should cover all accessible areas of the properties. Focus should be placed on operations primarily involving waste and materials handling practices. The presence of the consultant investigation team will raise questions regarding its purpose. Sensitivity issues are of major importance during the inspection of commercial properties with multiple tenants. The tenant areas must be inspected without causing alarm. Therefore, a good practice is to have the consultant accompanied by a representative of the client or owner, usually the property manager or maintenance supervisor, who is familiar with the tenant and their operations and who can answer questions asked by the building occupants. In addition, all requests for information concerning tenant operations should be directed through the property manager, owner, or client's designated representative. By following this protocol, sufficient information about tenant operations can be obtained to evaluate environmental conditions at the properties while minimizing any disturbance to tenant operations.

1.4.6 Obtaining Sufficient Information

Obtaining sufficient information in a timely manner so as not to impact the schedule is a major issue in performing the environmental assessment. A letter request for information should be sent to the property owner's contact person at the initiation of the assessment to allow as much time as possible for this information to be gathered. Depending on the availability and volume of information, this information can either be reviewed at the property or sent to the consultant's office for review. At least two different sources of information should be used, if possible, to verify a potential contamination problem during the Phase I process. If sufficient and reliable information cannot be provided to assess an environmental condition, recommendations should be provided to the client regarding the necessity for additional information and/or the need for an on-site investigation which would involve the collection and analysis of environmental samples. This work would be performed under the Phase II and, if necessary, the Phase III ESI process. All parties involved in the environmental assessment process should be made aware of the necessity of the requirement for adequate information to evaluate environmental conditions of the property.

REFERENCES

American Society of Testing Materials (ASTM), Subcommittee on Environmental Assessments for Commercial Real Estate. Standard E.50.02.2. *Standard Practice for Environmental Site Assessments: Phase I Environmental Site Assessment Process*, Working document, September 15, 1992.

Association of Groundwater Scientists and Engineers—Division of National Groundwater Association. *Guidance to Environmental Site Assessments*, September, 1992.

ASTM Subcommittee on Environmental Assessments for Commercial Real Estate. Standard E.50.02.1. *Standard Practice for Environmental Site Assessments: Transaction Screen Process*, Working Document, September 15, 1992.

Blumenfeld, K. 1982. *Remarks at the Environmental Liability Management Conference*, Crystal City, Virginia.

Buss, D.F., 1992. *Logistics and Management of Multifacility Environmental Assessments throughout the United States*, 47th Annual Industrial Waste Conference, Purdue University, Lafayette, Indiana.

Comprehensive Environmental Response, Compensation, and Liability Act of 1980, Section 101(35)(b), 42 U.S.C. Section 9601(35)(B) *et seq.*

Michigan Department of Natural Resources, 1983. *Minutes, Environmental Auditing Task Force*, State Bar Association of Michigan, Lansing, Michigan, April 11, 1983.

Tanner, Director Michigan Department of Natural Resources, Memorandum, September 24, 1982.

Additional Reading:

Colten, Craig E., editor, *Site Histories: Documenting Hazards for Environmental Site Assessments*, Cahners Publishing Company, Des Plaines, Illinois, 1993.

ESA Model Report: Model Phase I Environmental Site Assessment Report for Compliance with ASTM E 1527–93, ASFE, Silver Spring, Maryland, 1994.

Hess, Kathleen, *Environmental Site Assessment, Phase I: A Basic Guide*, Lewis Publishers, Boca Raton, Florida, 1993.

Phase I Environmental Site Assessments: The State of the Practice, ASFE, Silver Spring, Maryland, 1994.

Preacquisition Site Assessments: Recommended Management Procedures for Consulting Engineering Firms, ASFE. Silver Spring, Maryland, 1989.

Chapter 2

PHASE I—PRELIMINARY SITE ASSESSMENT

2.1 INTRODUCTION

As discussed in the first chapter of this manual, environmental liabilities in real estate transactions became an issue with the enactment of the Comprehensive Environmental Response Compensation, and Liability Act (CERCLA) of 1980 and the Superfund Reauthorization and Amendments Act of 1986 (SARA). The environmental consulting industry was created as a result to help assess and determine the nature and extent of these liabilities.

CERCLA imposes joint and several liability on potentially responsible parties (PRPs) without regard to fault for all costs associated with cleaning up contaminated properties. PRPs include, but are not limited to:

- the owner or operator of the property,
- the owner or operator of the property at the time the hazardous substances were disposed of,
- any person who arranged (by contract, agreement, or otherwise) for the disposal or treatment, or arranged with a transporter for transport for disposal or treatment of the hazardous substances, and
- any person who accepts hazardous substances for transport to disposal or treatment facilities.

These PRPs may be liable for (and potentially not limited to) the following costs:

- all costs of removal or remedial action incurred by the U.S. Government or a state,
- any other necessary costs of response incurred by any other person consistent with the National Contingency Plan,

- damages for injury to, destruction of, or loss of natural resources, including the cost of assessing the costs of these damages, and
- costs of any required health assessment or health effects study.

However, CERCLA provides four opportunities for minimizing liabilities in real estate transactions. Section 107(b) of the act allows a defendant to minimize or escape liability, known as the "Third Party Defense," if they can establish "by a preponderance of the evidence" that the release or threatened release of a hazardous substance and resulting damages were caused solely by:

- an act of God,
- an act of war,
- an act or omission of a third party other than an employee or agent of the defendant, or than one whose act or omission occurs in connection with a contractual relationship, existing directly or indirectly with the defendant.., if the defendant establishes that (a) he exercised due care with respect to the hazardous substance and (b) he took precautions against foreseeable acts or omissions of any such third party and the consequences that could foreseeably result from such acts or omissions, and
- any combination of the foregoing paragraphs.

To qualify for this defense, the defendant must show, by a preponderance of the evidence, that:

- the release or threat of release was caused solely by an act or omission of a third party other than an employee or agent of the defendant, or one whose act or omission occurs in connection with a contractual relationship, existing directly or indirectly, with the defendant,
- he exercised due care with respect to the hazardous substances concerned, taking into consideration the characteristics of such hazardous substance, in light of all relevant facts and circumstances, and
- he took precautions against foreseeable acts or omissions of any such third party and the consequences that could foreseeably result from such acts or omissions.

With the passage of SARA, other provisions for minimizing liability were made available:

- at the time the defendant acquired the facility the defendant did not know and had no reason to know that any hazardous substance which is the subject of the release or threatened release was disposed of on, in, or at the facility,
- the defendant is a government entity that acquired the facility by escheat, or through any other involuntary transfer or acquisition, or

through the exercise of eminent domain authority by purchase or condemnation, and
- the defendant acquired the facility by inheritance or bequest.

However, and this is the point from which environmental site assessments were created, to establish that the defendant had no reason to know, the defendant must have undertaken, at the time of acquisition, an appropriate inquiry into the previous ownership and uses of the property consistent with good commercial or customary practice in an effort to minimize liability.

In June 1989, EPA released a memorandum providing guidance on landowner liability under section 107(a)(1) of CERCLA, which specified the following information to be provided to establish the innocent landowner defense:

- evidence relevant to the actual or constructive knowledge of the landowner at the time of acquisition,
- affirmative steps taken by the landowner to determine previous ownership and uses of the property,
- condition of the property at the time of transfer,
- representations made at the time of transfer,
- purchase price of the property,
- fair market value of comparable property at the time of transfer, and
- any specialized knowledge of the land owner.

It is within the context of the "third party" and "innocent landowner" defenses, provided for in CERCLA and SARA, that this chapter examines and provides recommended practices for conducting a Phase I Preliminary Site Assessment (PSA).

The process described herein takes a "greatest common denominator" approach to investigating properties. It does not differentiate between property types, e.g., agricultural, or mixed residential commercial, or heavy industrial with site history, agency review, and site reconnaissance practices specifically targeted toward a certain property type. Rather, it emphasizes a more complete site characterization process, regardless of property type. Too often in the past, site assessors have assumed, based on property type, that certain records or information sources applied only to that type of property. This chapter proposes that those assumptions may be incorrect. Emphasis is placed on completing the most thorough site history, agency review, and site reconnaissance that can be accomplished based on project budgets and time constraints.

A suggested reading list pertaining to Environmental Site Assessments (ESAs) is included at the end of this chapter. It includes books, monographs, reports, and articles. Acquisition of these basic documents will be the foundation of a complete ESA library.

2.2 PRE-ASSESSMENT PLANNING

The success of a Phase I PSA is dependent on good planning and execution. Good planning begins with the initial client contact and extends to the point when the Phase I ESA is conducted. The planning process begins with the initial client contact and consists of, but is not limited to, the following basic questions to be answered.

Site Specific Questions

- What is the physical address of the site?
- Is it located within a city or is it within an unincorporated jurisdiction?
- What are the site boundaries—is a site map available?
- Is a legal description of the property available?
- Who are the neighbors/adjacent properties—to get a cursory understanding of adjacent land use?

Players

- Who is the client for billing purposes?
- Who is the current site owner?
- Who is the current site occupant?
- What is the caller's role in this process?
- What law firm, if any, represents the client?

Site History

- What were the known past uses of the site and surrounding area?
- Are there previous geotechnical or environmental studies of the site available?
- Are there any known past environmental problems?
- Are there any known past or present health or safety concerns or violations?

2.3 SITE OWNERSHIP AND LOCATION

It is important that the ownership and location of the site under study be identified before the PSA begins. Proper identification assures the client that the correct site is being investigated. Street addresses cannot be relied upon as the sole site location identifier. While the address may be accurate, there is no way of knowing the true boundaries of the site. Accordingly, the current legal description of the site should be obtained. This can usually be obtained from the client, a title company, or other sources if the tax number for the site has been identified. Also, a current plat map of the site boundaries should be obtained. This too is often available from the client or from the county or local tax assessment office. The plat should be confirmed as recorded and not in the approval process. The legal description obtained will be based on one of two methods: metes and bounds or plats and platting.

A metes and bounds survey is most common to rural and undeveloped areas, and consists of a walking tour of a property which begins by identifying an intersection of a government survey line. From that point, the survey describes boundary lines, giving directions in compass degrees and the distance before reaching the next corner of the property. This description continues around the property boundary until the survey starting point is reached. It is also common for rural properties to be described by a description of natural objects, such as streams and outcroppings, as long as the objects and boundaries are adequately described.

The system of plats and platting is most common to residential areas. Whenever property is divided into five or more lots or parcels for sale or lease, a plat map must be recorded with the local recorder's office. This map is a set of metes and bounds descriptions drawn on a map with reference to government survey lines. Once the plat has been signed and notarized it is recorded by the local property auditor in a book of plats.

2.4 SITE HISTORY EVALUATION

The site history evaluation is the first and most important element of the Phase I PSA. A comprehensive investigation of past and current uses of a site can help to determine the source, extent, and constituents of contaminants that may be present in a site's soil and groundwater or in buildings and structures. A good site history evaluation accomplishes the following.

- It characterizes known activity on and adjacent to a site from its undeveloped state to its current condition.
- It identifies past and current land use activities that are likely to have generated hazardous wastes of concern.
- It identifies potential contaminants released to the environment as a result of past and current land use activities.
- It assists engineers, geologists, and other related technical professionals in determining where to place soil borings and install monitoring wells for chemical sampling and analysis based on the location of historic activities of concern on a site.
- It is needed to fulfill the "due diligence" and "all appropriate inquiry" requirements of CERCLA and SARA.
- It helps identify possible adjacent properties that may have had releases that could migrate to the property being evaluated and be misinterpreted as having come from the site at issue.

2.4.1 Data Sources

To accomplish a good site history evaluation, a wide variety of data sources needs to be reviewed. The extent to which these data sources can be reviewed depends in part on the project budget and schedule. There are a variety of data sources that contain information relating to past and current site uses. These include aerial photographs, maps, real estate atlases, title documents, city street directories, building permits, textual records, historical records, and oral information. It is recommended that the consultant devise a checklist of site history investigation data sources so that all appropriate data sources are consulted if available. Specific descriptions of these categories of data sources follows. It is important to be able to gauge the importance of data sources based on the type of property being investigated. For instance, a rural or agricultural property requires different data sources than does a commercial property in a city's central business district. On the other hand, the time it takes to research information on a rural property, where land use documentation is limited, may equal or exceed the time required to assess the commercial property.

2.4.2 Aerial Photographs

Aerial photographs are an essential tool in determining past and current property uses. Figure 2–1 shows changing land usage along a shoreline. To accurately determine past and current land use activities of a site, however, all available aerial photographs should be acquired, if possible. A current aerial photograph is inadequate to document past site uses. In addition, aerial photographic interpretation may require more advanced skills than these needed for analyzing maps and real estate atlases. While most people have little trouble identifying land use activities in oblique aerial photographs (photographs taken at a low altitude and from an angle), vertical or near vertical aerial photography, the type most often used in site history evaluations, can be difficult to interpret and analyze. Accordingly, in order to accurately interpret vertical aerial photography, the investigator must be able to evaluate the following factors.

- *Shape*: Analysis of shape will distinguish between natural and man-made objects and structures. Shape interpretation can also identify unnatural topographic features.
- *Size*: Although a function of photographic scale, relative and absolute sizes are important in determining the dimensions of areas or activities of concern on a site.
- *Tone*: Features of different color have different qualities of light reflectance and thus, register varying shades or tones.
- *Shadow*: Due to the distortion present in truly vertical aerial photographs, shadows often reveal the purpose of natural or man-made objects.

Figure 2.1—Aerial photograph of the Seattle waterfront. (Washington
Department of Natural Resources, 1978)

- *Pattern*: Spatial arrangement differentiates between different types of land uses. For instance, a fruit orchard has a pattern different from a corn field.

To accurately interpret land use information in aerial photographs, the following equipment is recommended.

- Stereoscope. The function of this instrument is to deflect normally converging lines of sight so that each eye views a different photographic image. There are three kinds of stereoscopes: lens, mirror or reflecting, and zoom type.
- Stereometer or parallax bar for measuring heights.
- Engineer's scale, for measuring dimensions of natural and man made objects as well as possible waste spills.

- Tracing paper or vellum, to transfer land use information for ease of use as well as for use in preparing site maps.

In analyzing aerial photographs, look for the following:

- buildings and structures,
- aboveground storage tanks,
- roads and trails,
- ditches, pits, excavations, and lagoons,
- landfills,
- agricultural practices, such as farming and orcharding,
- waterbodies,
- evidence of possible hazardous substance dumping, and
- distressed vegetation.

Aerial photographs can be obtained from many sources. The availability of aerial photographs varies across the United States. It is important to be able to acquire photographs at a reasonable cost and within a short time frame. Readily available aerial photographs can be obtained from the following sources.

2.4.2.1 *Federal Government Sources*

Agricultural Stabilization & Conservation Service (ASCS)   The ASCS has used aerial photography in every state as the basic tool to determine acreage for farm program compliance purposes since 1935. This accounts for nearly 80% of the nation's total land area. Aerial photography flown for use by the ASCS, the Forest Service, and the Soil Conservation Service is on file at the ASCS's Aerial Photography Field Office located in Salt Lake City, Utah. However, aerial photography taken before 1941 has been transferred to the National Archives' Cartographic Archives Division. The ASCS also retains aerial photography taken as part of the National High Altitude Photography Program that began in 1978.

U.S. Geological Survey (USGS)   The USGS operates the National Aerial Photography Program (NAPP), a federal and state multi-agency activity that has developed a national data base of aerial photography availability. An important part of the NAPP is the Aerial Photography Summary Record System (APSRS) which offers a simple and inexpensive way to determine not only whether aerial coverage is available over a particular area, but whose photographic project it was, the conditions of the particular fly over, kind of photographic technique used, and who retains the film.

National Oceanic & Atmospheric Administration (NOAA)   NOAA aerial photography is secondary to NOAA's National Ocean Service activity of creating nautical charts. Nevertheless, NOAA can be an excellent source of aerial photography—provided the property under investigation is situated in a coastal environment.

Soil Conservation Service (SCS)   Normally, aerial photographs acquired by SCS offices are sent to the ASCS after completion of SCS projects, such as the preparation of a county soil survey, and generally 3-4 years after the flying date. Many SCS offices retain aerial photographs after project completion.

National Archives & Records Administration (NARA)   The archives' Cartographic and Architectural Branch has custody of some nine million aerial photographic images that were produced by both civilian and military agencies. Civilian agency photography is indexed by photomosaics by county and date from 1935 to 1945. Military agency photography is indexed by a series of 1:250,000 flight-line overlays that are arranged by degrees of latitude and longitude.

2.4.2.2 State and Local Government Sources

Department of Natural Resources   Most states have a department that manages natural resources such as forests, wetlands, aquatic lands, etc. Accordingly, they have aerial photography taken for inventory and planning purposes. As such they are an excellent source of photographs for site investigation purposes. Generally, these photographs are available for less than those offered by commercial vendors.

Department of Transportation   Every state has a department that manages state maintained trails, roads, and highways. Accordingly, they conduct fly-overs of major arterials on a regular basis as well as for areas in which transportation projects are planned. These will only be useful if the property under investigation is located near a state maintained highway.

2.4.2.3 Other Sources

Commercial Vendor Sources   Aerial photographs can be obtained from a wide variety of commercial vendors. These range from small private studios with stock aerial photography to larger laboratories that conduct aerial fly-overs for specific properties. Some of these vendors include those under contract to local, state, and federal agencies.

2.4.3 Maps

Maps are an important source of site history data because they may contain land use information. The following classes of maps are useful in determining past and current uses of a site:

- topographic,
- thematic (geologic, hydrologic, soil, climatologic, railroad, highway, forestry, and land use),
- historic, and
- planimetric.

Topographic maps are perhaps the most useful for site history evaluations because they show land uses such as structures, utility corridors, road types, and ground cover characteristics. In addition, topographic maps portray the shape and elevation of terrain using contours, form lines, shading, and color. The best known topographic maps are the quadrangle series maps published by the USGS. Topographic maps are classified by several publication scales—each serving a particular need. Map scales define the relationship between the measurement of map features and as they exist on the earth's surface. Thus, there are large scale maps (1:24,000), intermediate scale maps (1:50,000 to 1:100,000), and small scale maps (1:250,000 to 1:100,000,000). For site history evaluation purposes, the large scale map is the most useful because it provides the most detail. Other types of topographic maps include engineering maps, flood control maps published by the Federal Emergency Management Agency (FEMA), landscape maps, and bathymetric maps published by NOAA showing water depths and topography.

Thematic maps emphasize and devote the entire map to a particular topic such as geology, hydrology, soils, or land use. There are two classes of thematic maps: choropleth maps in which sections determined by civil boundaries or other divisions are colored, shaded, dotted, or in some way marked to show the proportion of the density of a given map's subject distribution, and isopleth maps showing numerical values of continuous distributions rather than discrete variables. For site history evaluation purposes, thematic maps of value include geologic maps indicating land uses, especially mining activity, land use maps which classify land use by category (industrial, commercial, residential, rural), soil maps portraying soil types and conditions, and flood insurance maps outlining 100-year flood boundaries. Flood insurance maps are particularly valuable because they serve as excellent city and town maps showing exact street locations and names.

Historic maps may be considered thematic; however, only those intended to show statistical information are classified as thematic. The most important historic map for site history evaluation purposes is the fire insurance map. Most common among the fire insurance map makers was the Sanborn Map Company of Pelham, New York. Between 1867 and the early 1950s, the company prepared detailed maps of most cities in the United States for the purpose of establishing fire insurance premium rates (Figures 2–2 to 2–3). Accordingly, these maps are perhaps the best historical record of actual land use activities on a site. For most large cities, these maps were prepared every few years and acquiring a complete set will provide detailed land use information of changes over time. Even for small cities and towns where only one map was published, it is information that should not be passed by during the site history evaluation. These maps are available on microfilm at many university and public libraries and from the Library of Congress.

Planimetric maps present the horizontal position of specific features but do not show relief in measurable forms. Examples of these maps include base, cadastral, line route, and outline maps. Base maps are used as a beginning point for creation of more specialized maps. Cadastral maps provide boundaries of subdivisions of land for the purpose of recording and describing ownership. The most common type of cadastral map is the plat which is often an essential part of a parcel of land's legal description. Utility companies are the primary users of line route maps that show the routes of pipelines, power lines, and other utility corridors. Outline maps are the most basic of planimetric maps, providing essential information to which additional information is compiled. In general, planimetric maps are less useful than other maps because they relate less to actual land use information and more to property boundaries. Nevertheless, they can be useful for establishing property boundaries and locations.

2.4.4 Real Estate Atlases

Real estate atlases can be considered a type of thematic map and are generally published for use in recording property ownership at the city and county level. The best known real estate atlas makers are the Metsker Map and Kroll Map companies, both located in Washington state. However, there are similar companies in the midwest and on the east coast. Generally, real estate atlases only show the name of property owners according to lot or parcel number. Sometimes the atlases will also include tax account numbers and structural outlines. For site history evaluation purposes, the city real estate atlas is more useful than the county atlas because it is more site specific. Like the fire insurance map, obtaining a succession of city real estate atlases will provide accurate site ownership (although not necessarily site utilization) information over time.

2.4.5 Title Searches and Documents

A traditional chain of title search examines the title of the owner of real estate to determine whether it is free and clear of encumbrances, and consists of an examination of such public records relating to or affecting real estate as established and maintained under authority of law. Unfortunately, for site history evaluation purposes, the traditional chain of title search is of little value because it only provides the sellers' names and any outstanding encumbrances or liens. Thus, a new type of chain of title search has emerged out of the need to know about past site uses. This new type of chain of title search is sometimes called a property title history report. When ordering a chain of title search, the consultant must specify how far back the search should go. At this time, it is generally accepted to conduct a 50 year search. However, for sites in commercial or industrial areas or where land use is known to have existed beyond 50 years it may be prudent to have a 75 or even 100 year search conducted. The following documents should be requested.

- Deed. A written instrument conveying certain rights in real property. The written instrument may be a quit claim deed, trust deed, or warranty deed.

SANBORN MAP LEGEND

CODING OF NON-RESIDENTIAL FIRE-RESISTIVE STRUCTURAL UNITS FOR FIREPROOF AND NON-COMBUSTIBLE BUILDINGS

FRAMING

CODE	STRUCTURAL UNIT
A.	Reinforced Concrete Frame.
B.	Reinforced Concrete Joists, Columns, Beams, Trusses, Arches, Masonry Piers.
C.	Protected Steel Frame.
D.	Individually Protected Steel Joists, Columns, Beams, Trusses, Arches.
E.	Indirectly Protected Steel Frame.
F.	Indirectly Protected Steel Joists, Columns, Beams, Trusses, Arches.
G.	Unprotected Steel Joists, Columns, Beams, Trusses, Arches.
H.	Unprotected Steel Frame.
O.	Masonry Bearing Walls only.

FLOORS

CODE	STRUCTURAL UNIT
1.	Reinforced Concrete, Reinforced Concrete with Masonry Units, Pre-cast Concrete or Gypsum Slabs or Planks.
2.	Concrete on Metal Lath, Incombustible Form Boards, Paper-backed Wire Fabric, Steel Deck, or Cellular, Ribbed or Corrugated Steel Units.
3.	Open Steel Deck or Grating.

ROOF

CODE	STRUCTURAL UNIT
a.	Reinforced Concrete, Reinforced Concrete with Masonry Units, Reinforced Gypsum Concrete, Pre-cast Concrete or Gypsum Slabs or Planks.
b.	Concrete or Gypsum on Metal Lath, Incombustible Form Boards, Paper-backed Wire Fabric, Steel Deck, or Cellular, Ribbed or Corrugated Steel Units.
c.	Incombustible Composition Boards with or without Insulation, Masonry or Metal Tiles.
d.	Steel Deck, Corrugated Metal or Asbestos Protected Metal with or without Insulation.

The coding to left, for framing, floor and roof structural units is used in describing the construction of fire-resistive buildings. In addition, wall construction other than brick, will show above the date built, wall construction other than brick, and ceilings.

FP – 1962 (CONC.) A – 1 – a — A fireproof building built in 1962 with concrete walls and reinforced concrete frame, floors and roof.

FPX – 1962 (METAL PANELS) B-2-a — A fireproof building built in 1962 with metal panel walls, reinforced concrete column and beams, concrete floors on metal lath and gypsum slab roof, non-combustible ceilings.

NC -1962 (C.B.) H- 2 - d — A noncombustible building built in 1962 with concrete block walls; unprotected steel columns, beams and joists; concrete floors on metal lath and steel deck roof.

MASONRY CONSTRUCTION

Important interior and all exterior masonry walls of all non-residential buildings are shown with weighted (——) lines. Masonry walls of all residential buildings are shown with a standard line and the construction is noted on all buildings diagrammed after July, 1963.

WALLS

- 8" Brick
- 12" Concrete
- 18" & 20" Stone
- 12" & 8" Hollow Tile, Wall Thicknesses Placed Relative to Respective Floors
- Cinder, Concrete or Cement Block
- Hollow Cinder or Concrete Block Interior Wall Basement to Roof
- Tile Interior Wall Basement to Roof
- Cement Brick End Wall

- Mixed Construction of Concrete Blocks, Brick Faced
- Mixed Construction of Concrete Blocks & Brick
- Masonry Walls, Metal Faced
- Adobe
- Hollow Cinder or Concrete Block 1st Floor only
- Brick 2nd Floor only
- Tile 1st & 3rd Floors only

PARTITIONS

- Frame
- Tile from Foundation to Top Ceiling only
- Concrete 1st Floor only
- Hollow Cinder or Concrete Block 1st Floor only
- Brick 2nd Floor only
- Tile 1st & 3rd Floors only

OPENINGS (Interior)

- Wall with No Openings
- Wall with Double Standard Fire Doors 1st Floor
- Wall with Standard Fire Doors 1st & 2nd Floor Basement
- Wall with Substandard Fire Doors 1st & 3rd Floors
- Wall with Metal & Wired Glass Fire Doors all Floors
- Wall with Substandard Fire Doors 1st, 2nd & 3rd Floors & Unprotected Opening 4th Floor
- Wall with Small Unprotected Openings 4th Floor
- Wall with Unprotected Openings all Floors

OPENINGS (Exterior)

- 1st Floor
- 1st & 2nd Floors
- 3rd Floor
- 1st & 4th Fl. with Metal Shutter 1st.
- 10th & 22nd only
- 10th to 22nd Fl.
- Glass Block
- Wired Glass in Metal Sash 2nd & 3rd Fl.

NON-MASONRY CONSTRUCTION

Non-masonry walls are shown with fine (——) lines. (Wall construction other than wood and stucco on wood frame is noted)

- Wood & Stucco & Cement Plaster, Etc. on Wood Frame
- Brick Veneered on Wood Frame (Other type of Veneer on Wood Frame Specifically Noted)
- Mixed Masonry & Non-Masonry (Type of Masonry Specifically Noted)
- Wood, Brick Lined, Br. Filled or Brick Nogged

- Iron Building with Wood Roof, (Location of Exterior Wood Areas Specifically noted)
- Asbestos Clad on Wood Frame (Used in All Residential Structures only.)
- Mixed Wall—9" of CB With Metal Sash Above
- Metal Panels

- Wood Sash & Glass
- Metal Sash & Glass
- Metal Clad on Wood
- Iron Building

- Apron Walls With Wood Sash and Glass
- Stucco, Cement Plaster, Etc. on Steel Frame
- Gunite on Steel Frame

- Asphalt and/or Asbestos Protected Metal on Steel Frame
- Asphalt and/or Asbestos Protected Metal on Wood Frame
- Glass Panels

GLOSSARY

A – B LINES An arbitrary boundary between adjoining sheets.
A.F. Private garage.
ABV. Above.
A.D.T. Equipped with fire detecting device which automatically registers reports of central fire department.
AIR COND. A cooling system employing ducts through floors.
APRON WALL A masonry wall not underwriting by Stock Fire Ins. Companies.
ASSOC. RISK Risk not underwritten by Stock Fire Ins. Companies.
BASEMENT A story having the floor below ground & its ceiling at least 4' above ground.
Cook County, Ill.: A floor of a building next below the first floor. Shown by the symbol SB following the story height. Sub-basements or sub-cellars, (stories below the 1st basement), are shown by the symbol SSB following the basement symbol. Mountain & Pacific Coast States: Basements & cellars.
CHIMNEYS (Applicable to maps in Rocky Mountain & Pacific Coast States)
B.C. Brick, stone, concrete brick & concrete chimney.
C.BL.C. Concrete block chimney.
N.C. Non-standard concrete chimney.
T.C. Tile chimney.
P.C. Patent chimney.
BR. Brick chimney.
IR. Iron chimney.
S.P. Stove pipe.
S.P.V. Stove pipe with patent ventilator.

RESIDENTIAL OCCUPANCY SYMBOLS

S. Single family unit or as qualified by a number.
F.... APTS. A multi-family residential building corresponding with local Building Bureau definition in family units per floor, story height, & separation of entrances.
ROOMS. A residential building normally occupied by a single family but with 10 or more rooms rented for lodging purposes.
EXCEPTIONS. 6 rooms in Arizona, California, Nevada, Utah, Montana, 5 rooms in Oregon & Washington, 4 rooms in Idaho & Hawaii.

FIRE RESISTIVE CONSTRUCTION SYMBOLS

F.P. Approved masonry walls, floors & roofs of approved masonry, concrete, and/or protected steel.
F.P.X. F.P. qualifications, except interior or sub-standard walls.
N.C. Fire resistive with unprotected structural steel units.
HOLLOW WALL A bonded masonry wall having a continuous air space within.
I.F. Independent Electric Plant.
IMPASSABLE Not traversable due to condition of terrain.
LEDGED WALL A masonry bearing wall with extended ledges to support floors.
LOFT Tenanted by industrial occupancies.
N.G. Concrete or plaster applied to metal lath on wood studdings.
NR. Metal sash & glass.
NO OPGS. Streets appearing on records but not open on ground.
O.H. Windows overlooking the roof above the corresponding floor of an adjoining building.
O.L.U. Open between ground and first floor.
PLASTO. Masonry reinforcing columns on diagram.
SKY. Skylights.
SL. Slate attached to wood siding.
SM.HO. Smoke House.
SUSP. Suspended ceilings below floor for air conditioning.
SYST. System.
TRANS. Transformer.
WD. Wood.

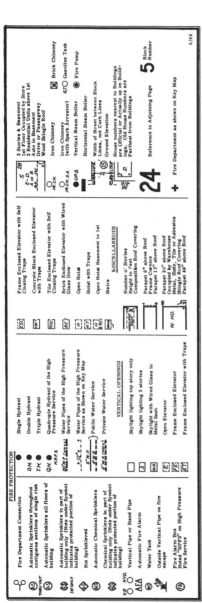

FIRE PROTECTION

Fire Department Connection	Single Hydrant
Automatic Sprinklers throughout, contiguous sections of single risk	Double Hydrant
Automatic Sprinklers all floors of building	Triple Hydrant
Automatic Sprinklers in part of building only (Note under Symbol indicate protected portion of building)	Quadruple Hydrant of the High Pressure Service
Not Sprinklered	Water Pipes of the High Pressure Service
Automatic Chemical Sprinklers	Water Pipes of the High Pressure Service as Shown on Key Map
Chemical Sprinklers in part of building only (Note under Symbol indicate protected portion of building)	Public Water Service
	Private Water Service

VERTICAL OPENINGS

Vertical Pipe or Stand Pipe	Skylight lighting top story only	Frame Enclosed Elevator with Self Closing Traps
Automatic Fire Alarm	Skylight lighting 3 stories	Concrete Block Enclosed Elevator with Traps
Water Tank	Skylight with Wired Glass in Metal Sash	Tile Enclosed Elevator with Self Closing Traps
Outside Vertical Pipe on fire escape	Open Elevator	Brick Enclosed Elevator with Wired Glass Door
Fire Alarm Box Noted "HPFS" on High Pressure Fire Service	Frame Enclosed Elevator	Open Hoist
	Frame Enclosed Elevator with Traps	Hoist with Traps
		Open Hoist Basement to 1st
		Stairs

MISCELLANEOUS

- Number o' Stories
- Height in Feet
- Composition Roof Covering
- Parapet 6" above Roof
- Frame Cornice
- Parapet 13" above Roof
- Parapet 24" above Roof Occupied by Warehouse
- Metal, Slate, Tile or Asbestos Shingle Roof Covering
- Parapet 44" above Roof

- 2 Stories & Basement 1st Floor Occupied by Store 2 Residential Units above 1st Auto in Basement Drive or Passageway Wood Shingle Roof
- Brick Chimney
- Iron Chimney
- Iron Chimney (with Spark Arrestor) 67 Gasoline Tank
- Vertical Steam Boiler ● Fire Pump
- Horizontal Steam Boiler
- Width of Street between Block Lines, not Curb Lines
- Ground Elevation
- House numbers nearest to Buildings are Official or Actually up on Buildings. Old House Numbers are Farthest from Buildings
- Reference to Adjoining Page 5 Block Number
- Fire Department as shown on Key Map

6/69

Map legend for Large and small sized B & W Sanborn Map Editions.*
published after 1950 by Sanborn Mapping and Geographic Information Service.
COPYRIGHT © 1995, SANBORN MAPPING AND GEOGRAHIC INFORMATION SERVICE

* B & W Editions ' B Maps ' are those maps that are made from B & W mylar masters, with all information conveyed through the use of alpha-numeric symbols and shading, not B & W reproductions of color masters-with the traditional paste up corrections attached to the maps.

Figure 2.2—(continued)

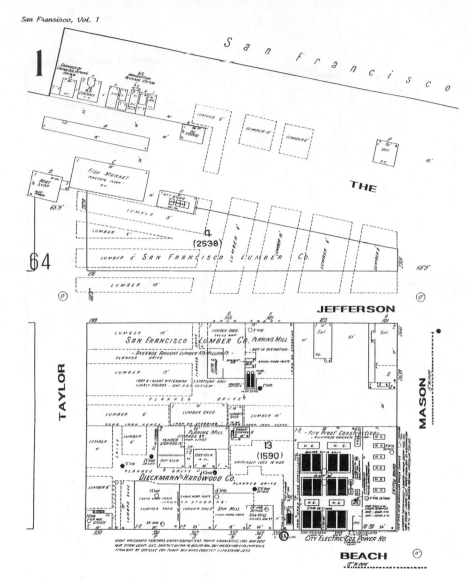

Figure 2.3—Sanborn Map of the Fishermans Terminal area of San Francisco in 1913 and 1990

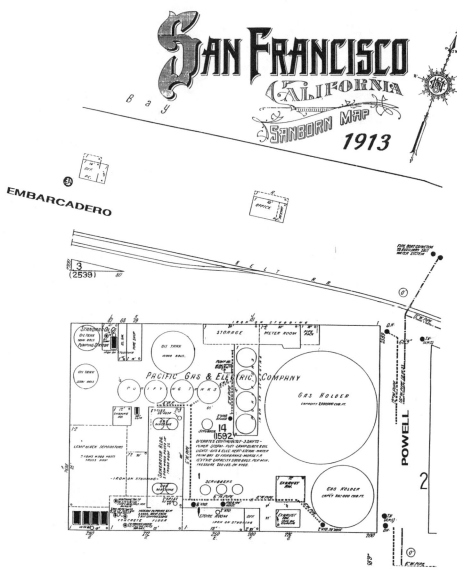

Figure 2.3—(continued)

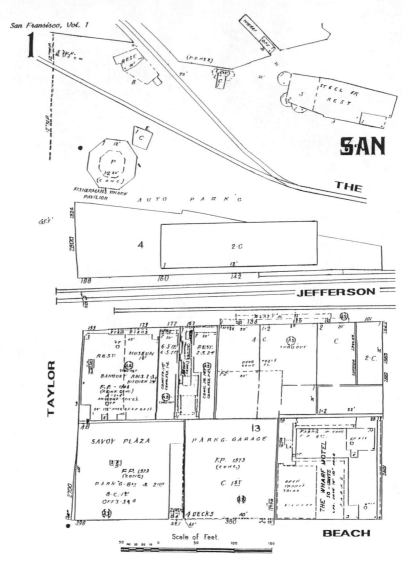

Figure 2.3—(continued)

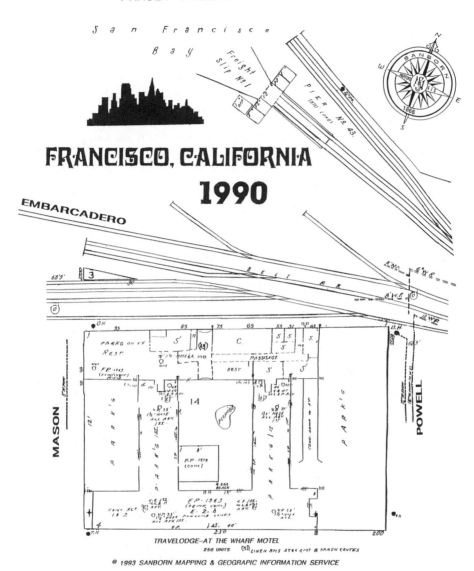

@ 1993 SANBORN MAPPING & GEOGRAPIC INFORMATION SERVICE

Figure 2.3—(continued)

- Easement. The right to use or enjoy certain privileges that appertain to the land of another, such as a right-of-way by reason of an agreement of record with the property owner.
- Lease. A contract between an owner of real estate and a tenant stating the considerations and conditions upon which the tenant may occupy and use the property, as well as the responsibilities of the owner and tenant.
- Lien. A hold or claim that one person has upon the property of another as a security for some debt or charge. Environmental liens explain why a lender has placed a lien on a property as a result of environmental problems.

The following items should be requested when ordering a chain of title search: 1) a typed chronological listing of all known real estate transactions related to the site, 2) photocopies of all title documents including deeds, easements, leases, and environmental liens, and 3) a map showing the location of the site (many title companies provide a location map as it is sometimes the only good site map one will receive before beginning a site history evaluation).

2.4.6 City Street Directories

Street directories exist for most first, second, and third class cities in the United States. R.L. Polk & Company based in Kansas City, Missouri is the best known publisher of such directories. These directories may be found at the local library of the subject city. They are very valuable for site history evaluations because when researched over time they provide exact information on who has lived at, or operated a business at, a certain street address. However, these directories are most useful from about 1939 to the present. Between the 1880s and the late 1930s, the directories only listed street addresses by business type. On or about 1939, the publisher began dividing the directory into three parts: 1) alphabetic listing of individuals by name, 2) chronological listing of addresses by named street, and 3) chronological listing of street addresses by numbered street. These are commonly known as reverse directories because they allow the researcher to reverse the normal address research process of searching by name.

2.4.7 Building Permits and Plans

An excellent source of historic site use information are building permits and building plans. These are generally available from city or county building departments. Building permits and plans can provide land use information on a property over time. Generally, building permits are issued for the following activities: building construction and remodeling, mechanical, grading, and heating and ventilation systems. These permits may also include certificates of occupancy and land use files that contain information on short plats, lot boundary adjustments, rezoning, conditional uses, variances, and environmental impact assessments and statements. Building permits often include information on the installation of underground storage tanks and the installation of equipment known to generate hazardous substances such as spray paint booths, vapor degreasers, septic tanks, and oil fired kilns. Building plans often accompany building permits.

These plans may have information on construction materials of concern. Many large cities have microfilmed plans for commercial and multifamily buildings constructed since the turn of the century. Few cities, however, keep plans for single family homes less current than five or ten years.

2.4.8 Textual Records

Textual records of value for site history evaluation purposes include books, reports, monographs, dissertations, theses, newspapers, and periodicals (magazines and journals). However, some textual records of interest that are not historical in nature but may prove useful in determining past site uses may include such topics as natural resources, urban planning, industrial technology, and geography. Textual records on industrial technology are particularly helpful in determining how certain industrial waste was managed and how processing operations have changed over time. Textual records are available in city, county, regional, university, state, and federal agency libraries.

2.4.9 Archival Records and Manuscripts

Although these are the least likely data sources to be used in a site history evaluation, they should not be overlooked if budgets and schedules allow. Archival records are generally records of an institution, agency, or organization maintained by that entity. Manuscripts consist of business records and personal papers collected by museums, historical societies, and university libraries for the purpose of research. For site history evaluation purposes, business records will prove the most valuable because they contain records relating to business operations and administration. Archival records and manuscripts can be found at the local, state, and federal level. Descriptions of some of these data sources are presented below.

- Historical Societies can be an important source of property use information. Every state, and many cities and counties have their own historical society. There are also many thematic societies, such as those devoted to agriculture, aviation, industry, and science. Historical societies are useful because they contain local history information in books, reports, dissertations, theses, newspapers, magazines, photographs, and historic records of businesses and citizens. Historical societies also are an excellent source of personal knowledge in the form of docents (volunteers) and society members. Generally, these

people are the older citizens of the area and often have an encyclopedic knowledge of land use activities in a certain area.

- A museum is an institution that preserves and exhibits artistic, historical, or scientific objects. As such, many museums at the local,
state, and federal level exhibit historical objects relating to property
uses in the form of maps, aerial photographs, dioramas, and film. In
addition, many museums maintain historical records for research
purposes—especially state and county museums and thematic historical museums (railroading, aviation, etc.).
- University libraries can be an excellent source of local history information. Resources typically available in these repositories include
aerial photographs, maps, atlases, newspapers, magazines and periodicals, and textual records in the form of books, reports, dissertations, and theses.
- Although their availability to the public is limited, corporate archives can be an excellent source of property use information. For
instance, if a site history evaluation is being conducted of an oil
terminal, shipyard, or pulp and paper mill, records of operation and
processing changes over time may be available in the corporate
archives. These may also contain aerial photographs and video tapes.
- Most large cities and many large counties have archives that reflect
city and county development. Information useful for site history
evaluation are records on public works and utilities construction,
aerial and subject photographs, maps, drawings, and building permits.
- Every state acquires, arranges, describes, preserves, and makes available to the public, the historically valuable records of state government. These records can be an excellent source of property use
information, especially records of agencies that were involved in
natural resource and land use regulation, zoning, tax assessment,
and environmental management. Every state has a library maintained for the purpose of acquiring all state government publications
as well as serving as a research repository for the state's citizens.
Older agency reports, especially those related to natural resources
and land use can prove valuable for site history evaluation purposes.
In some cases, state libraries also maintain their own archives.
- Founded in 1938, the National Archives and Records Administration
(NARA) acquires, arranges, describes, preserves, and makes available to the public the historically valuable records of the three
branches of the government: over 3 billion textual documents, 2
million cartographic items, 9 million aerial photographs, 91 million
feet of motion picture film, and 120,000 videotapes and sound recordings. In addition, NARA provides the same service for records
transferred to twelve regional archives. For site history evaluation
purposes, these will prove the most useful because they directly
relate to specific regions and states. NARA also manages seven presi-
dential libraries, each of which contains the personal papers of the

president, legislators, cabinet members, associates, friends, and family. These can be a good source of property information.

- Oral documentation can be an effective method of identifying past uses of a property. At a minimum, oral interviews should be conducted with the current site owners and current site occupants. Often, the site owner is not the actual site operator (for instance when the owner has leased or rented out the property to another party). It can also be advantageous to conduct oral interviews with former site owners and occupants. However, with the increasing public knowledge of environmental liabilities, past owners and users may not be willing to talk to site assessors.

- Given the viability and reliability of an individual's verbal information, care must be taken to confirm or verify that information by reliable written documentation. If the only source of information was through an oral interview, then the use of that information in the report should be so noted. This can be done in the text as, "Smith, J., verbal communication, 1995," and, if important enough, within the Limitations section of the report.

2.5 SITE RECONNAISSANCE

The purpose of the site reconnaissance is to evaluate current land use and environmental conditions to determine the potential for contaminants to exist in the property's soil and groundwater based on current on-site and adjacent site uses. This is achieved through a combination of visual and photographic documentation, written documentation, and personal interviews.

2.5.1 Visual and Photographic Documentation

It is important that all areas and items of concern on and adjacent to the site be documented. This is achieved by a combination of visual observations and photographic documentation. One way to approach the site reconnaissance is to systematically walk over the site in a grid pattern, i.e., establish a walking line north to south and then east to west. In this way, the entire site will be covered. However, if the site contains buildings and structures, it may be necessary to divide the site into zones.

Items or areas of concern on the site must be documented, at a minimum, with photographs. This shows that an effort has been made to document concerns and that the site was actually visited by the consultant. If possible, use a camera that dates the film. This will carry more weight in court should the need arise to prove that site reconnaissance was conducted. In some cases, it may be prudent to video tape the site reconnaissance with the owner's approval. This can prove useful for very large properties as well as for sites that are known to have items and areas of concern.

It is usually not possible to conduct as thorough a reconnaissance of adjacent properties because of limited access or because site occupants or owners will not allow such activity to occur. In these cases, it is important to document adjacent land use and environmental conditions by camera or video camera. However, there may be times when the site owner will not allow photographic documentation, and should be so noted in the site reconnaissance field notes. Even more care must be taken in preparing a site map based solely on visual observations.

2.5.2 Written Documentation

In addition to visual and photographic documentation of the site reconnaissance, it is important to document observations of site and adjacent land use and environmental conditions. It is recommended that a site reconnaissance checklist be used as a guide in reconnoitering the site. A variety of such checklists have been developed by lending institutions, banks, law firms, real estate firms, and regulatory agencies. As part of the written documentation it may be necessary to sketch site features and items of interest for use later in preparing the site map for the Phase I PSA report.

2.5.3 Personal Interviews

Available on-site and adjacent land occupants or owner/operators should be interviewed regarding their knowledge of past and contemporary uses of the site and surrounding area. Questions should reflect the information needed from the categories presented in the site reconnaissance checklist. Since oral information can be considered hearsay in court, it is important to corroborate oral information with written documentation if at all possible. Also, it is important to be able to judge the credibility and level of knowledge of the interviewee. Accordingly, shop foreman, plant managers, and other personnel in a supervisory position should be interviewed first. Copies of all available pertinent environmental permits, inspections reports, and notices of violations and/or penalties should also be obtained.

2.5.4 Site and Environmental Characterization

Specific items to look for during a site reconnaissance can be divided into two distinct categories. The first category is to characterize land uses of the site; the second is to characterize environmental conditions of the site. Although distinct, any opportunity to observe items of concern in the land use characterization is essential. Specific items or conditions to look for are described below.

2.5.4.1 Land Use Characterization

Physical Features and Surface Conditions

Parking, Roads, and Trails   Accessibility to the site is important from the standpoint of who is able to gain access to the site and the activities that may result from that access. Generally, commercial and industrial sites have well controlled access, but rural properties do not. The ability for (and signs of) young children or adolescents to gain access to a site, especially if dangerous conditions exist on-site is a critical piece of information. Accordingly, it is important to look for evidence of improper access or activities, such as dumping, that may imply improper access.

Fences and Gates   This is ancillary to the above category. Fences often denote property boundaries so it is important to note the presence of such features in the site reconnaissance.

Loading Zones   Many commercial and industrial sites, in particular warehouses or facilities where goods and materials are received and shipped, have loading areas where hydraulically operated lifts are or previously were located. The major concern here is the presence of leaking hydraulic fluids, as well as the presence of petroleum products leaked from trucks during the loading and off-loading process or even from accidental spills. It is common for small spills to go unreported.

Depressions   A hollow area completely surrounded by higher ground and having no natural outlet should be observed for containing any hazardous substances or materials of concern. Depressions are often used on rural and undeveloped properties for illegal dumping activity.

Vegetative Distress   A variety of activities can cause vegetative distress and frequently this is a sign of surface contamination. Vegetative distress includes dead vegetation caused by herbicide application or the spillage or leakage of hazardous materials. Also, discoloration can occur from chemical application and spills. Another form of vegetative distress is that caused by ground clearing activities or the placement of heavy equipment on the ground.

Stained Soil   Soil that has been stained by petroleum products and other chemicals or materials is quite common on commercial and industrial properties. It is important to note the location and extent of the stain(s).

Erosion   Erosion on a site can also be caused by man-made activities rather than through natural processes. This is important because non-natural erosion may impact the site (particularly sites located along water bodies or wetlands) by allowing waste streams to impact areas that would have otherwise not been affected.

Rock Outcroppings   Surficially exposed bedrock at or near the site can provide useful information about bedrock conditions, such as fracturing, rock type, permeability, and possible depths of soil.

Filling   The presence of fill material should be noted because it is often used to cover up evidence of hazardous substance spills and disposal areas. Fill material is frequently used to even out unusual topographic features. The quantity, quality, and source of any on-site fill should be of concern. Fill material can include demolition and construction debris, trash and garbage, and hazardous substances.

Surface Water Conditions

Drainage   The process of water discharge from, to, or through a site is of interest because it is a potential receptor of hazardous substances. Without actually sampling surface water, it is not possible to determine actual environmental conditions. However, evidence of oil or other chemical sheens or the presence of strong chemical odors are a sign of potential problems.

Waterbodies   Waterbodies refer to rivers, lakes, streams, or ponds that either exist in their entirety on a site or for which a portion is on a site. These are of interest because of their potential as a receptor of hazardous substances. In addition, it is common for water bodies to contain submerged hazardous substances and materials.

Wetlands   Wetlands are a potential receptor of hazardous substances. In addition, their presence on or near a site may impact future use of the site due to waterfowl and wildlife that may inhabit the wetland.

Utilities

Power Poles and Transformers   It is important to identify the location of power poles on a site because they often include transformers which are a potential source of polychlorinated biphenyls (PCBs). The power pole should be examined to see if the electrical company or other owner has identified the transformer as having been tested for the presence of PCBs. If there is no such indication, then note the power pole number and the transformer number, and contact the owner of the power pole to determine its PCB status. It is also important to note if the transformer has leaked onto the pole or down to the ground surface. Most power companies do not test their transformers on a regular basis. Rather, testing occurs when the pole is taken out of service or needs repair. Most often, the power company will charge the site owner a small fee to test for PCBs and dispose of the transformer.

Water lines   If a developed site has no on-site well(s) then it is likely to be tied into the local water system. Water lines feeding the site could represent pathways of migration for hazardous fumes or for spilled liquid substances and need to be identified.

Sewer lines   Similarly, if a site has no on-site septic system then it is likely hooked into a local sewer system. For the same reasons as water lines, sewer lines need to be noted.

Outfalls   Sites located along a marine environment or along surface water bodies may have drainage outfalls. These systems also represent possible pathways for contaminants and, where they discharge, may show visible signs of possible hazardous substances (vegetative stress, odors, stains, etc.).

Catch Basins and Storm Drains   Most commercial and industrial sites have more than one storm drain or catch basin. The environmental professional should determine whether the storm drain or catch basin discharges to the ground, waterbody, or local sewer system and is unobstructed. He should also look for evidence of oily or other chemical sheens and chemical odors. For sites where oily water discharges are permitted, a functioning oil/water separator should be installed in the storm drain or catch basin. Backups may be an indication of poor housekeeping practices on the part of the facility or may indicate a problem with the system.

Septic Systems and Drain Fields   These systems typically consist of a settling tank in which settled sludge is in contact with sewage flowing through the tank while the sludge is being decomposed by anaerobic bacterial action. Septic systems are most common in rural and residential settings. However, some commercial and industrial facilities have their wastewater discharged to a local sewer system and then have their domestic waste discharged to a septic system. In some cases, hazardous substances are discharged to septic systems (evidence of this may appear as unusual odors, or leachate in the system's drain field), or solvent-based chemicals have been used to clean out the system and/or connecting lines. Vegetative stress in the vicinity of the leachfield may also be an indicator of the presence of or a former release of hazardous substances.

Wells   Wells (deep production-type or monitoring) can now be found on most properties. In observing wells on a site, note the type of well, its location, and its construction characteristics. It is also important to note whether the well cap has been tampered with, because it is common to dump hazardous substances into wells. Another type of well is a dry well—typically shallow (no more than 15-20 feet deep), fairly large in diameter (between 25-35 inches), and filled with crushed rock or similar material. They are commonly and usually used for infiltration of storm water runoff. However, it is common to find dry wells on commercial and industrial properties that have been used for disposal of liquid hazardous materials.

Buildings and Structures
Name and Nature of Business   The environmental professional should accurately record the name of the business (if any) located at the site. Sometimes a client will provide one name, when the actual name is different. In addition, site historical information can be obtained when older business names are noted on buildings and structures. This is especially true for commercial and industrial facilities that have had many different owners or have changed their scope of manufacturing over time.

Physical Location   Accurately record where the building(s) and/or structure(s) are located on the site. Orientation is important, particularly when determining where soil sampling or groundwater wells should be placed if additional investigation is warranted.

Age The age of a facility is important from a site history standpoint. If a facility is older than forty to fifty years, it is possible that the facility changed some aspects of its processing and manufacturing activities over time, and may have used certain materials that are no longer allowed (e.g., asbestos insulation).

Dimensions As part of the overall process of understanding site uses, the dimensions of all on-site buildings and structures should be obtained.

Construction Type The type of construction used in buildings and structures should be noted. This is important because many materials used in construction contain substances that may present a threat to human health and the environment. This can range from creosote coated or soaked railroad ties used in slope stabilization to the use of floor tiles and other building products that contain asbestos.

Asbestos Evaluating a site for the possible presence of asbestos can only be done in a cursory manner, i.e. noting the age of a building, insulation type, etc. Trained and certified individuals are now required in may states to make these observations and to conduct sampling and testing. Asbestos was historically used in a variety of ways in buildings and structures. For instance, asbestos can be found in floor tiles, ceiling tiles, some paints, pipe wrapping, insulation, and drywall. These are commonly referred to as ACMs, asbestos containing materials.

PCBs Polychlorinated biphenyls (PCBs) are most often found in electrical transformers. Transformers can be found on power poles and as stand alone units. Regardless of their location or size, transformers should be checked for evidence, such as a sticker or placard, that the local power company or other jurisdiction has tested it for the presence of PCBs.

Urea Formaldehyde Foam Insulation (UFFI) A known carcinogen, urea insulation is no longer used. However, it may be found in older buildings and structures. Again, for the Phase I PSA only a cursory evaluation need be accomplished. In fact it may be virtually impossible for the investigator to identify this material as it may not be exposed.

Lead Paint Lead paint is very common in buildings and structures constructed prior to 1979. The presence of lead paint cannot be determined without physical testing. However, painted buildings and structures, predating 1979 may be assumed to contain such paint.

Utilities Just as various utilities are noted on the site surface, so should utilities be noted inside buildings and structures. The existence of the following utilities should be noted: electrical, natural gas, propane, water, sewer, and oil. Many of these lines may represent a potential pathway of contaminant migration or a potential point of release of materials.

Floor Drains The presence of floor drains in a building or structure indicates that some type of liquid is currently or was previously discharged. Most often, drains are used to collect precipitation runoff through bay or garage doors or for process water. However, in commercial and industrial facilities, hazardous substances and chemicals are sometimes directed toward floor drains. The drains should be inspected for possible chemical odors and petroleum and chemical sheens.

Sumps   A sump is a tank that receives and temporarily stores liquids at the lowest point of a circulating or drainage system. They can sometimes contain hazardous substances or oily discharges. Sumps are often located in generating rooms and power stations to which blowoff water is discharged.

Spill Evidence   Throughout the site reconnaissance, every effort should be made to note any sign of spillage on the ground, into conveyance structures and in buildings and other site structures. Evidence of spills shows that in the past some activity has taken place which may result in site contamination.

Adjacent Properties   As part of the site reconnaissance, it is important to observe, to the extent possible, land use characteristics and environmental conditions of immediately adjacent properties. This is important because of the potential contamination that has been released on that property to migrate onto the property you are evaluating. In observing adjacent properties, the items previously described should be evaluated and include the following:

- physical features and surface conditions,
- surface water conditions,
- utilities, and
- buildings/structures.

2.5.4.2 *Environmental Characterization*

Evidence of Hazardous Substances

Drums and Barrels   Drums and barrels are common signs that a property or facility stores or in some other way manages liquids or solids. An inventory of the drums and barrels observed during a site reconnaissance should be made. Their condition, identifying marks, and contents, if determinable, should be noted. If they are rusted or have leaked, there is a potential for soil and groundwater contamination. Be aware that a label does not necessarily confirm what is in that drum or barrel. If possible, verify label contents with a visual inspection. Safety considerations may severely limit the investigators' visual inspection of drum contents.

Discolored Surface Water   Surface water that is discolored, as may be the case with petroleum contamination, may indicate a disposal problem nearby or where the surface water originates.

Stained Soil   Stained soil often represents the release of a substance or chemical that may be hazardous. Often, stained soil occurs as a result of accidental spillage or leakage of petroleum products. This can occur through simple tasks such as oil changes and vehicle maintenance, but also by spillage or leakage from containers.

Vegetative Distress   A variety of activities can cause vegetative distress and frequently this is a sign of surface or near surface contamination. Vegetative distress is often caused by herbicide application, the spillage or leakage of hazardous substances, or discoloration from chemical application. Another form of vegetative distress is that caused by ground clearing activities or the placement of heavy equipment or objects.

Waste Management Practices

Generator Identification Number   Most facilities or businesses that have been inspected or in some other way monitored by the EPA or its state equivalent may have been assigned a unique number. Firms generating a certain amount of hazardous waste will have applied for and have a hazardous waste generator identification number. Obtain this number and, as available, acquire the following related documents from the key site manager. If not available on-site, these may be acquired through a file review at the responsible agency.

- *Compliance Inspections*: Conducted by federal, state, and local officials, compliance inspections provide information on a facility's compliance with federal, state, and local environmental regulations. These inspections will have information on the state of the facility as of the day of the inspection. These are a "snap shot" in time and are valuable for providing historical compliance information on a facility. These inspections may take the form of a checklist and narrative, and often include photographs, maps, drawings, and plans.
- *Spill Reports*: If there has been a spill at the facility, then the manager should have a copy of the spill report. This will provide information on the cause of the spill and how it was managed. These reports may also include photographs and other information.
- *Permits*: Most commercial and industrial facilities have one or more permits required by federal, state, or local agencies. These permits/licenses should be available from the facility manager. Having a permit does not necessarily imply a lack of environmental problems. But the fact that a permit has been granted shows that the facility is involved in some way with the management of a hazardous substance or chemical that is of enough interest to a regulatory agency that the facility or operation needed a permit.
- *Penalties and Violations*: These records may be difficult to obtain during a site reconnaissance as the facility may want to put on the "best face" possible. The presence of such documents means that there has been an inappropriate use of or management of hazardous substances or chemicals. Such records should be available from regulatory agencies.
- *Chemical Inventory*: Most facilities keep track of the chemicals they use and obtaining this inventory will provide information on the types of chemicals used.

- *Material Safety Data Sheets*: These data sheets are required to be maintained at the site with the facility's chemical inventory. They are important because they provide toxicological information about chemicals used or stored at the site.

Waste Disposal Methods
Storage   Hazardous and/or other wastes are often stored on-site. Note the type of storage medium (closet, bin, drums and barrels, cabinets, etc.) and whether or not the storage area has a secondary containment system to prevent spills or leaks from spreading to other areas. To the extent possible, determine whether the containment area has been compromised. In addition, take an inventory of the waste contained in the storage area(s). Many facilities are required to maintain a Spill Prevention, Control, and Countermeasure (SPCC) Plan which should detail their measures to contain, control, and/or prevent spills.

Location   Inspect and evaluate the location of all stored waste for any signs of spills or leaks.

Processing Operations
Product(s)   Knowing the product that results from a facility's processing operation will provide clues as to the processes and chemicals used in creating the product. This can be helpful if no listing or written documentation is available.

Process(es)   Critical to the understanding of a manufacturing facility are the processing operations used at the facility. Knowing how a product is made provides a better understanding of the environmental liabilities that exist in a facility.

Chemicals Utilized   As part of the processing reconnaissance, note the types of chemicals used, how often they are used in the process, and how they are managed.

Spill Evidence   Note whether or not the processing operations have resulted in the release of chemicals. For example, in a metal fabrication shop, it is important to note if press brakes and other fabricating machines have leaked hydraulic fluids.

Underground and Aboveground Storage Tanks (UST/AST)
Location   Identify where all known UST/AST(s) are located on a property.

Capacity   To the extent possible, determine the dimensions and capacity of each UST/AST and determine the quantity of substance in the UST/AST.

Use Status   To the extent possible, determine whether or not the UST/AST is currently being used.

Spill or Leak Evidence   It is important to note any evidence of spillage or leakage around an UST/AST. Fortunately for the investigator, petroleum staining is fairly persistent in the environment and will remain visible for a long time after the spill or leak event. Although minor in quantity, it is very common for petroleum products to be spilled during tank filling.

Leak Detection or Spill Prevention   Determine whether or not each tank has leak detection or spill detection monitoring systems in place, or a spill containment system.

Age   If possible, determine the age of the UST/AST. This is important because older tanks are more likely to have leaked over time.

Construction   For ASTs, note the type of construction and whether the tank's structural integrity has been compromised by weathering, poor construction, or improper use. Also verify the condition of the associated piping and the use of any containment system for the piping.

Adjacent Properties   As part of the site reconnaissance, it is important to observe, to the extent possible, environmental conditions of immediately adjacent properties. This is important because of the potential for migratory contamination. In observing adjacent properties, the items previously described should be evaluated and include the following:

- evidence of hazardous substances,
- waste management practices,
- waste disposal methods,
- processing operations, and
- UST/AST.

2.6 REGULATORY INFORMATION REVIEW

The acquisition and analysis of available local (city, county, regional governments), state, and federal information related to known environmental conditions of a site and other sites of concern within a designated radius of the site is a critical step in determining the potential for contaminants of concern to exist on a site. Information obtained in this review may also be valuable to the site history evaluation because local, state, and federal data sources often include historic site use information. A thorough local, state, and federal regulatory information review accomplishes the following.

- It identifies known environmental problems associated with a site and its surrounding area.
- It describes available documented information related to a site and surrounding area relative to the potential for contaminants to exist on the site due to past practices or to have migrated to the site due to past practices at other sites.

It is recommended that a checklist be developed to assist the consultant in ensuring that all appropriate local, state, and federal agencies are contacted for information on the site and surrounding area. To perform a thorough local, state, and federal information review, a wide variety of data sources need to be reviewed. These data sources are available in the form of reports, lists, data bases, and original agency files. A description of suggested sources to consult is presented in the following sections.

Before commencing the regulatory information review, some thought should be given to the size of the area around a site for which other sites of concern should be researched. The ASTM standard for Phase I PSAs established a set radii (called the approximate minimum search distance) for certain types of regulatory information (called the standard environmental record source). There is some controversy over the suggested ASTM radii, primarily related to the one-mile radius established for searching certain regulatory information. Adherence to the ASTM standard is acceptable when only abstracting information from reports, lists, and databases. However, adherence to the standard becomes burdensome when researching listed sites' original agency files. For instance, researching available agency files for all sites within a one-mile radius of a rural or agricultural property may not be prohibitive. However, for a highly industrialized area, such research may be very extensive and, therefore, cost or time prohibitive. Accordingly, it may be more appropriate to limit the agency file search to sites located between one-quarter and one-half mile of the subject site. Where appropriate, some exceptions should be made. All state and federal Superfund site files within a mile of a subject site should be reviewed. In addition, agency files of upgradient industrial or commercial sites, regardless of the radii, should be investigated if there is a potential for contaminant migration to the subject site. However, it may not be necessary to review all agency files of treatment, storage, and disposal facilities regulated by the Resource Conservation and Recovery Act, as amended (RCRA) and other sites within a one-mile radius of the subject site. It may also be unnecessary to review all agency files of sites within a one-half mile radius of the subject site that appear on the Comprehensive Environmental Response, Compensation, & Liability Information System (CERCLIS), state landfill, or UST lists.

2.6.1 Federal Data Sources

U.S. Environmental Protection Agency (EPA): The EPA is the federal agency with information most relevant to the Phase I ESA. Information of value can be found in the form of reports, lists, data bases, and files maintained by EPA in its regional offices. The Facilities Index Data System (FINDS) is a database maintained by EPA that contains entries for more than 450,000 facilities regulated by EPA under various enforcement programs as well as enforcement programs run by the states. FINDS can be considered the "master index" of sites for which environmental monitoring is on-going. However, it lists only basic enforcement data for each site or facility. Hence, in many cases it is important that the database of origin be perused so that all appropriate information is obtained. Nevertheless, FINDS is an excellent starting point for determining whether or not a site and its surrounding area has been, or is under, investigation by a regional EPA office. Each entry includes the name of the site, a variety of geographic locators (street address, city, county, zipcode, latitude and longitude, congressional district), codes identifying the site, and a list of other EPA and state databases containing information about the site, including the code numbers used to retrieve pertinent records from other databases. The FINDS database contains entries for sites/facilities covered in the following separate EPA databases.

- *AIRS*: The Aerometric Information Retrieval System is maintained by EPA's Office of Air & Radiation and contains entries on sites or facilities subject to air quality monitoring or reporting under the Clean Air Act.
- *CERCLIS*: The Comprehensive Environmental Response Cleanup and Liability Information System is a database used by EPA to track activities conducted under its Superfund program. Specific information is tracked for each individual site. Sites that come to EPA's attention that may have a potential for releasing hazardous substances into the environment are added to the inventory. EPA learns of these sites from owner notification, citizen complaints, state and local government identification and as a result of other EPA investigations.

There are three categories of sites on the CERCLIS inventory: sites that may be potentially hazardous and require preliminary investigation, sites for which no further remedial action is planned, and sites that have been investigated and which EPA has determined may represent a long term threat to public health and/or the environment.

It is important to know that the CERCLIS inventory is often used as a resource in environmental site investigations and that just because a site is contained in CERCLIS does not necessarily mean it is contaminated. Conversely, the absence of a site does not necessarily mean it is contaminant free. The CERCLIS inventory is available at all EPA regional offices as well as through private sources.

A subset of CERCLIS is the National Priorities List (NPL), which is a list of sites on CERCLIS that have been investigated by the EPA and determined to be among the most hazardous of all sites in the United States.

- *CICIS*: The Chemicals in Commerce Information System is maintained by EPA's Office of Pesticide and Toxic Substances and contains information about manufacturers of substances listed on the Toxic Substances Control Act (TSCA) inventory of hazardous substances.
- *CUS*: The Chemical Update System is maintained by EPA's Office of Pesticides and Toxic Substances and contains references to facilities that manufactured or imported more than 10,000 pounds of specific chemicals during the preceding year.
- *DOCKET*: The Civil Enforcement Docket is maintained by EPA's Office of Enforcement and Compliance Monitoring and contains information on civil and judicial cases filed by the justice department on behalf of EPA.

- *FATES*: The Federal Insecticide, Fungicide, and Rodenticide Act (FIFRA) and Toxic Substances Control Act (TSCA) Enforcement System is maintained by EPA's Office of Prevention, Pesticides and Toxic Substances and contains references to inspections of sites or facilities under the jurisdiction of FIFRA.
- *FFIS*: The Federal Facilities Information System is maintained by EPA's Office of Federal Affairs and contains information on all projects undertaken by federal agencies at federal facilities to maintain or achieve compliance with environmental laws and regulations.
- *PADS*: The PCBs Activity Data System is maintained by the Chemical Regulations Branch of EPA's Office of Pesticides and Toxic Substances and contains records of sites or facilities that transport, store, or dispose of PCBs.
- *PCS*: The Permit Compliance System is maintained by EPA's Office of Water Enforcement and Permits, and is concerned with large scale waste water treatment plants.
- *RCRAJ*: This database is maintained by EPA's Office of Solid Waste and was established under authority of the Medical Waste Tracking Act of 1988, which added subtitle J to RCRA. It contains information related to haulers of medical waste in Rhode Island, Connecticut, New York, and New Jersey.
- *RCRIS*: The RCRA Information System is maintained by EPA's Office of Solid Waste. It contains entries for every site or facility regulated under RCRA. These sites are generally involved in the generation, transport, disposal, or storing of hazardous materials.
- *TRIS*: The Toxics Release Inventory System is maintained by the Office of Pesticide and Toxic Substances and contains information pertaining to releases of certain toxic substances as required under Title III of SARA. This database is available at very little cost through the National Library of Medicine's MEDLARS system.

In addition to EPA databases noted in FINDS, it is also important to review available agency files available in the EPA regional office that covers the area in which the project is located. In most cases, it is necessary to file a Freedom of Information Act (FOIA) request to review hard copy business specific files. Many EPA regional offices allow these requests to be filed by fax. Whatever the format received by EPA, under current law, EPA must respond to a FOIA request within ten days from the date it is received. Often this is not enough time to be of value. When requesting a "file review," a list of businesses, their street address and zip codes, and the type of files to be reviewed should be noted. All available files related to air, groundwater, surface water, and hazardous waste issues should be requested. Generally, each EPA regional office has a FOIA coordinator that will handle the FOIA request. They will route requests to the appropriate regional offices to determine whether or not any of the files exist. Then, if files are available, a file review appointment can be arranged. The file review will usually take place in an empty conference room or office. Photocopying fees as well as research time fees for those who pulled the records of interest may be assessed.

2.6.2 State and Local Data Sources

It is difficult to characterize the types of data sources available at the state and local level due to the number of states, counties, boroughs, parishes, cities, towns, municipalities, and regional governments. However, some generalizations can be made about the data sources that are available from these repositories.

2.6.2.1 State Agencies Most states have a department that is the equivalent of EPA, and maintain records that are similar in scope to those maintained by EPA. Most states also have FOIA type requirements. It is important to review all available hard copy files relating to businesses on and adjacent to the property under study. The available records may include the following:

- hazardous waste sites list,
- registered underground storage tank list (UST),
- leaking underground storage tank list (LUST), and
- solid waste landfill list.

In addition, it may be useful to acquire well logs for all wells within a prescribed radius of the property under study or, within a specific section of a township and range. Well logs provide data on soils encountered during drilling through a cross-section drawing. A rudimentary understanding of groundwater flow and depths can be abstracted from static water level averaging and topographic placement of the wells.

2.6.2.2 County and Local Government Agencies Valuable information on property uses can be obtained from a wide variety of county and other local government agencies. In general, the following agencies may prove useful in obtaining information about the property under study.

- *Planning Department*: The city, town, county, or municipal planning departments may have information on land use and facility design or expansion as a result of the preparation of environmental impact statements as well as other base-line studies. Information in the form of aerial photographs and land use maps is usually available.
- *Zoning Department*: This department provides information on allowable uses of properties. Past site use information can be obtained in archival records of these departments such as historic zoning maps and surveys.
- *Tax Assessor*: Information on property uses can be obtained from the assessor in the form of plats and plat books. In many cases, information on property characteristics, property ownership history, and property legal descriptions can be obtained instantly by computer.

Further, tax records many contain information on land improve-
ments and remodeling dates of buildings and structures. In many
states, tax assessor records are transferred to state or local archives,
and these can be excellent sources of historic land use information,
including real property assessment rolls, personal property assessment rolls, delinquent tax rolls, board of equalization and other
miscellaneous assessment records, and real property recordings.

- *Environmental Health Department*: This department will have information on such things as septic systems, land-filling, dumping complaints, and other issues impacting human health and the
  environment in the area of the department's jurisdiction. Often, the
  best source of information in these departments is the sanitary engineer or environmental inspector. Since they are often in the field, they
  may have encyclopedic knowledge of the local area based on their
  field work.
- *Fire Department/Fire District*: The fire department will have information on fire and hazardous material response calls to businesses. In
  some cases, the fire department also has jurisdiction over the installation and removal of underground and above ground storage tanks.
  These records may be available in hard copy and possibly on micro-
  fiche or microfilm.
- *Regional Pollution Control Agencies*: Many states now have regional
  pollution control authorities, particularly for air quality monitoring.
  These agencies will have comprehensive records relating to emissions from industry and commercial operations. Records will include
  inspection reports, permits, permit applications, notices of violations, and penalties assessed.
- *Public Works Department*: This department will have information on
  the construction of public works projects adjacent to properties under study.
- *Utilities Departments*: This department will have information on utilities on and adjacent to the property under study. Perhaps most
  important will be their records pertaining to the status of PCB oils in
  electrical power pole transformers.

2.6.3 Commercial Data Sources

There now are a variety of commercial vendors offering fee-based
agency information reviews based on searches of local, state, and federal
environmental databases. These searches are usually based on zipcodes
identified within the radius of the study area. These searches are an
option to consider for those who do not have easy access or the time
available to consult the aforementioned local, state, and federal agencies.
However, these vendors only offer database generated reports. Accordingly, an in-person visit must be made to the appropriate agencies to
research hard copy records, the background data from which database
entries are made. Some vendors also make available maps of all known

hazardous waste sites usually within a one-mile radius of the property under study. Typically, these vendors search the following federal databases: CERCLIS, NPL, RCRA, TRIS, TSCA, and SPILLS, and the following state databases: UST, LUST, hazardous waste sites, and landfills. Thus, information obtained from the vendors usually consists of two things: a printout containing the results of federal and state database searches, and a mapped presentation of sites identified in the federal and state databases.

2.7 PHASE I REPORT FORMAT

When the Phase I investigation is completed, the findings must be provided to the client in a clear and concise manner. Findings and, when necessary, recommendations for further investigation must be presented in language that a layperson will understand. The format presented below provides an excellent outline for preparing a Phase I report. This format allows for a structured, yet flexible, method of presenting the findings, opinions, and recommendations of the investigation.

2.7.1 Cover Letter

The purpose of the cover letter is to provide the client with a very brief overview of the results of the Phase I investigation. It should be no more than two pages long. Sometimes, this can be expanded into an executive summary; some clients specifically request an executive summary.

2.7.2 Table of Contents

This presents the order in which project information is presented in the report. The table of contents should include primary and secondary report headings as well as a list of appendices, figures, and tables.

2.7.3 Introduction

The introduction is to refamiliarize the reader with the purpose of the investigation and incorporates the statement of understanding from the proposal.

2.7.4 Scope of Services

This provides contractual details about the project that were agreed upon in the proposal. It outlines the scope of services proposed and performed.

2.7.5 Study Area Boundaries

It is important to inform the reader of the study area. Since the Phase I investigation examines land uses adjacent to the site, specific boundaries must be identified. Study area boundaries will range from immediately adjacent for the site history evaluation and site reconnaissance to a number of radii between one-quarter and one mile for the agency information review.

2.7.6 Site Description and Setting

This section lets the reader know that the consultant is familiar with site and surrounding area characteristics. It reaffirms that the consultant has investigated the correct property. This is the section of the report that presents a vicinity map showing the location of the site in relation to the surrounding area.

2.7.7 Site History Evaluation

This section examines past uses of the site and surrounding area to determine the potential for contaminants to exist on, or migrate from the site. This section can be divided into three parts. Part one describes the sources used in researching the site history, part two presents information on the historical uses of the site, and part three relates specifically to historical uses of adjacent properties.

2.7.8 Regulatory Information Review

This section should list all agencies contacted for information relating to the environmental quality of the site and adjacent properties. Agencies should be grouped according to their jurisdiction: city, county, regional, state, and federal. For each agency contacted, list the person(s) contacted, telephone numbers if known, and a description of information (whether database information or in-person file review) reviewed. Information pertinent to the site should be included if it will not be presented elsewhere.

2.7.9 Site Reconnaissance

Observations of current land use and environmental conditions on the site and adjacent to the site are described in this section. These should be based on visual observations (the site walkover), photographic documentation (camera or video camera), and written communication (checklist). This section should also include a site map which includes the site boundaries and all items of concern noted during the site reconnaissance.

2.7.10 Regulatory Issues

It is important to educate readers about regulatory issues involved in real estate transactions. This section should provide information on any local, state, or federal regulations that may impact the site under study. This could include areas or activities at the site that require permitting but have not been. Care must be taken as to what is stated in this section.

The Phase I report could be used as official notification of a violation by a regulating agency.

2.7.11 Summary of Findings and Opinions

This section reiterates the information presented in the body of the report. It should summarize, by sub-section, what was found through site history review, agency information review, and site reconnaissance. Then, it should describe the potential for contamination based on these reviews.

2.7.12 Recommendations

If it is determined that there is a potential for contamination to exist on the property based on past uses, documented environmental problems, or current land use conditions, it will be necessary to provide recommendations as appropriate, for further study. These recommendations can range from a simple statement that more study is needed to characterizing soil and groundwater conditions to a scope of services and cost estimate for a Phase II study.

2.7.13 Limitations of the Report

This section includes information to the effect that the report has been prepared for the exclusive use of the client, that the report is based on information obtained and conditions observed during the investigation and that no responsibility can be accepted for deficiencies, misstatements, omissions, misrepresentations, or the fraudulent acts of others, and that the services have been performed in general accordance with standards of the industry for similar projects.

Further detailing of any situations, such as the inability of the investigator to observe a certain area of the site or building due to lack of access, inability to acquire and review all relevant information due to time or access constraints, etc., should be listed.

2.7.14 References

As necessary, include all references cited in the text. This does not include all data sources used to characterize past uses of the site, only those sources, e.g., oral contacts, quotes from books, reports, and agency files, that are referenced in the report text.

2.7.15 Appendices

The content of report appendices will vary according to client demands and property types, but can include the following: chain of title search, site reconnaissance photographs, and agency information review sources of particular interest (notices of violations, inspections reports, etc.).

REFERENCES

ASTM. *ASTM Standards on Environmental Site Assessments for Commercial Real Estate*, Philadelphia, 1993.

Bureau of National Affairs. *Environmental Due Diligence Guide*, Washington, D.C., 1993.

Colangelo, Robert V. *Buyer Beware: The Fundamentals of Environmental Property Assessments*, National Groundwater Association, Dublin, Ohio, 1991.

Colten, C.E. *Site Histories: Documenting Hazards for Environmental Site Assessments*, Cahners Publishing Company, Des Plaines, Iowa, 1993.

Mulville-Friel Diane. *Guidelines and Methods for Conducting Property Transfer Site Histories*, Illinois State Museum, Springfield, 1990.

CECMW. *Environmental Site Assessments: Guidance for Users and Providers, The Mid Atlantic Standard of Care*, Silver Spring, Maryland, 1992.

Marburg Associates and William Parkin. *Site Auditing: Environmental Assessment of Property*, Specialty Technical Publishers, Inc., Vancouver, B.C., 1993.

Moskowitz, Joel S. *Environmental Liability and Real Property Transactions: Law and Practice*, Wiley Law Publications, New York, 1989.

Nanney, Donald C. *Environmental Risks in Real Estate Transactions*, McGraw-Hill, New York, 1993.

National Groundwater Association. *Proceedings of 1991 Environmental Site Assessments Case Studies and Strategies Conference*, Dublin, Ohio, 1991.

Guidance to Environmental Site Assessments, Dublin, Ohio, 1992.

Proceedings of Environmental Site Assessments Case Study and Strategies, The Conference, Dublin, Ohio, 1992.

Policy Planning & Evaluation, Inc. *Selected Current Practices in Property Transfer Environmental Assessment*, A Report Prepared for the U.S. EPA. Washington, D.C., 1989.

Prescott, Michael K. and Douglas S. Brossman. *The Environmental Handbook for Property Transfer and Financing*, Lewis Publishers, Chelsea, Michigan, 1990.

Shineldecker, Chris L. *Handbook of Environmental Contaminants: A Guide for Site Assessment*, Lewis Publishers, Chelsea, Michigan, 1992.

Stirling, Dale A. "Site Histories in Environmental Site Assessments: A New Opportunity for Public Historians." *The Public Historian*, 12(2):45–52, Spring 1990.

"Incorporating History into Environmental Site Assessments." *Environmental & Waste Management World*, 6(6):9–11, August 1992.

"Sanborn Fire Insurance Maps in Environmental Site Assessments." *Environmental & Waste Management World*, 7(1):8–12, January/February 1993.

"Site Assessors Warned Not to Rely on Standard Database Searches Alone." *HAZMAT World*, 7(4):54–56, April 1994.

"Environmental Site Assessments Face Evolving Standards." Parts I and II. *Washington Environmental Compliance Update*, 1(6), July 1994.

"Environmental Site Assessments Face Evolving Standards." Part II. *Washington Environmental Compliance Update*, 1(7), August 1994.

Tarantino, John A. *Environmental Liability Transaction Guide Forms and Checklists*, Wiley Law Publications, New York, 1992.

Thompson, Morris M. *Maps for America*, U.S. Geological Survey, 1981.

U.S. EPA. Guidance on Landowner Liability Under Section 107(a)(1) and De Minimis Settlements Under Section 122(g)(1)(B) of CERCLA, and Settlements with Prospective Purchasers of Contaminated Property. OSWER Directive 9835.9, 1989.

Chapter 3

PHASE II—SITE INVESTIGATION

3.1 INTRODUCTION

No two sites are alike, so a "cookbook" investigation, particularly in phases of investigation beyond Phase I assessment background research and site reconnaissance, is neither efficient or cost-effective. A phased approach is effective because the scope of each phase builds on information gathered in the previous phase, allowing the environmental consultant conducting the investigation to better identify and characterize potential contamination problems at the property. Other advantages of the phased approach include: prevention of costly "over-investigation," better control of information flow (more information is not always better if there is no strategy or plan for addressing the findings), and completion of a more meaningful and directed investigation tailored to the history of a property and the individual need of the sponsoring party.

Successful Phase I investigations are limited by the diligence of the investigator and how well public and facility files are maintained. In subsequent phases of investigation, the ground rules are more general in nature because objectives are project and site specific, there are more decision points in design and implementation of the investigation, and the reporting of results can be more subjective depending on the audience for whom the evaluation is completed. Simply stated, there is much more room for variability. Variability, coupled with the fact that society is becoming increasingly litigious, point to a glaring need for an established industry standard of care for conducting not only Phase I environmental site assessments, but Phase II–type investigations as well.

This chapter is not intended to define what should be a part of a specific Phase II investigation, but rather to serve as guidance or a starting point for design and implementation of any such investigation. Although the actual scope of Phase II investigations can vary widely, activities can generally be broken down into four basic stages: pre-investigation planning, data collection, data interpretation, and reporting. Figure 3–1 illustrates the Phase II investigation process by breaking down associated activities required to fulfill the needs of each of these stages. Sections of this chapter have been dedicated to more detailed description and discussion of the activities that might be included in each of these stages.

3.1.1 Definition of Phase II Investigation

Simply stated, a Phase II Investigation is an environmental site assessment that includes a subsurface investigation and/or sampling component whose major purpose is to characterize and confirm the presence of on-site contamination. In general, it includes invasive and non-invasive subsurface investigation, chemical analysis of air, soil, source material, surface water, and/or groundwater samples collected on-site, evaluation of the analytical and hydrogeological data, and, frequently, some form of risk assessment. In the context of this manual, a Phase II investigation is an investigation that builds on and confirms information identified during the Phase I assessment.

Environmental site assessment terminology can be confusing if taken out of context; for instance, what is referred to as a Phase II investigation in this manual may also be called a Phase I, if the initial phase is called a "Preliminary Site Assessment" or "PA." Likewise, various federal and state regulations may refer to the Phase II investigation as a "site inspection," "limited site investigation," Tier II investigation," "subsurface investigation," or just a plain "field investigation." The key identifier is the inclusion of sample collection and analytical data evaluation.

3.1.2 Purpose of Phase II Investigation

Depending on the intent or objective of the party performing or requiring the investigation and the results of the Phase I investigation, a Phase II may not always be necessary. The need for a Phase II investigation may be different for different parties. For example, all of the following parties might go forward with a Phase II: investors or purchasers (if financing requires tangible proof of "clean" property), a property owner (to document impact from an off-site source or to confirm no off-site impact from the property), a property owner or operator (to fulfill a regulatory requirement, to establish baseline environmental conditions, or to quantify environmental liability or risk to site users), and any party to a litigation matter (with the burden of proving a property is clean or without risk).

It is common for parties on opposite sides of a transaction to pass back and forth the responsibility for completing different phases of a site assessment. For instance, a seller may complete a Phase I to help the marketability of his property, but a potential buyer may have a need to confirm the results of the Phase I, or go one step further to document the paper trail by conducting a Phase II investigation. If contamination is

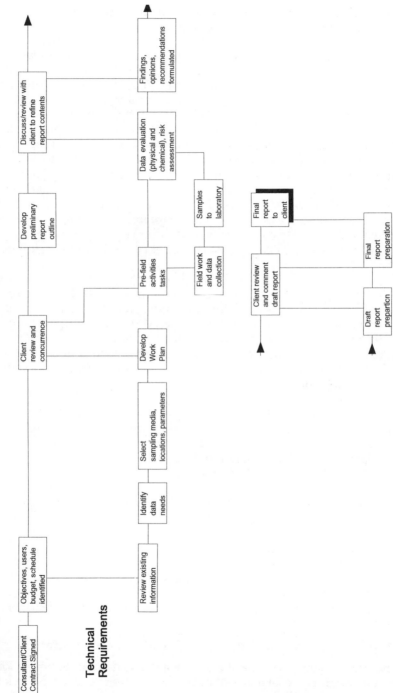

Figure 3.1—Phase II Investigation Process

found during Phase II, the responsibility for further delineation or risk characterization may again shift back to the seller.

Environmental liability issues drive the site assessment process. Outside of the regulatory arena, environmental site assessments are most often performed to satisfy transactional due diligence requirements when transferring (buying or selling), financing, investing, renting, abandoning rental or subletting real property or for corporate acquisition (assets purchase in particular). The decision to go forward always depends on the parties, their objectives, and/or their interest in the property (a measure of liability). For example, a corporate investor might not commission a Phase II if the Phase I provides enough information to quantify and/or make a determination regarding potential or actual environmental liability associated with a property. Simply summarized, the decision to go forward is made relative to individual circumstances. It is important, when reviewing a report completed for a different party, to understand the circumstance(s) and purpose(s) for which that party commissioned the study. Their interests may not be the same as the interested party.

The Phase I report is generally the starting point for identification of the technical issues suggesting the need for a Phase II based on the potential or actual release of contaminants. Indicators of this include: industrial site history or hazardous material use, documentation of on-site disposal or spills, an adjacent property with an apparent environmental problem such as a contaminated groundwater plume, history of regulatory enforcement for environmental violations, and prior sampling data indicating contamination.

A 1992 Survey conducted by the Consulting Engineers Council of Metropolitan Washington identified the most common factors that result in a Phase II investigation. They are as follows by percentage resulting from Phase I Site Assessment.

83%   Discovery of Underground Storage Tanks
66%   Past Land Use
46%   Use of Adjacent Property
27%   Client Request
26%   PCB Transformers
21%   Buildings Constructed Before 1975
23%   Other

3.1.3 Matching Purpose and Perspective with Level of Effort

There is no standard scope of work for a Phase II investigation. A Phase II investigation could include all of the components of a field investigation that are described in this chapter, or it could include only a few.

Likewise, costs for a Phase II investigation can vary from $10,000 to $200,000 with most in the $10,000 to $50,000 range. Costs also vary widely depending on the part of the country in which the site is located; the northeast and the west coast are generally more expensive. Definition of the level of effort required for a Phase II investigation is limited by two extremes: looking for anything that could be there by implementing a study that leaves no unanswered questions (an all-encompassing study— usually a regulator's approach), or implementing a study that focuses on only known problems, such as a spill area (a focused study, the approach of the regulated). Either approach would be appropriate under the right circumstances. However, most Phase II investigations fall somewhere in the middle because they are designed to serve multiple purposes (e.g. both transactional due diligence and regulatory compliance).

In summary, the purpose for which a Phase II investigation is performed dictates the appropriate level of work and confidence necessary to satisfy the end-user of the study or the sponsoring party. Phase II investigations are performed for transactional purposes [due diligence, litigation purposes, or regulatory purposes (enforcement or voluntary compliance)]. They may also be incident-focused (e.g., spill related). Each purpose has inherent requirements to be met. The scope or requirements may also vary depending on the perspective of the sponsoring party (same purpose, different perspective or interests). Experienced environmental professionals can help determine what level of effort is necessary for a set of circumstances. Likewise, an environmental attorney may also be consulted for more guidance or insight on this subject. It is important to recognize early on that why one is completing a Phase II investigation is as important as what will be included in the study. The two issues go hand in hand.

3.2 PRE-INVESTIGATION PLANNING

The key to providing for a technically well-run project and effective cost control lies in good up-front planning. The primary end-product resulting from the pre-investigation planning stage of a Phase II investigation is the written scope of work, proposal, or work plan, as the case may be. The development of these pre-investigation work products most frequently begins before a contract is signed since generally the scope and magnitude of services must be defined and agreed upon for the contract.

3.2.1 Pre-Investigation Information Gathering

The first contacts between the environmental consultant and the client should be handled in such a way as to begin this information gathering and goals definition process. Consideration of a number of questions by both parties can facilitate the process.

- Where is the site?
- Who owns it?

- What is the objective of the investigation?
- Who will use the investigation's results and for what purpose?
- How many parties are involved and what are their roles?
- Has other work already been completed?
- What are the schedule requirements and constraints?

3.2.1.1 *Identification of Objectives and Requirements*   Effective communication between the client and consultant is important in establishing clearly understood objectives and requirements. Teaching/ educating is part of a consultant's job; being a good listener is also important. Clients can also educate their consultants regarding their industry and site-specific interests, issues, and concerns. Project specific objectives and requirements should be refined through discussion and review with the client and then included in a written scope of work or other contractual document. The following is a list of points to reflect on when identifying the objectives and requirements of a specific Phase II investigation.

- *Client Needs*: A client's needs and requirements should be determined up-front (beginning with the first phone conversation) and should serve as a point of comparison for future project reviews.
- *Project Objectives*: The understanding of the investigation objectives should be refined through discussion. Restate objectives in writing in the proposal or written scope of work.
- *Report End-Users*: As an outgrowth of defining objectives, the end-users define the format for reporting of information.
- *Investigation Limitations*: All involved parties should have a clear understanding of what the proposed scope of services will and will not accomplish. Clients should be asked to carefully read the limitations that usually accompany a report of findings, and the assumptions that usually accompany a proposal. Clients should understand that investigative technology is limited. Likewise, consultants should explain such limitations up-front (before work begins).
- *Regulatory Requirements*: A client's intended use defines what regulatory requirements must be met. Or, in the case of a Phase II investigation conducted for other than regulatory purposes, it should be clearly stated that the investigation is *not* intended to meet specific regulatory requirements. A few states have regulatory requirements that a site assessment be conducted each time property ownership is transferred. A few more states have published specific guidance for performing Phase II investigations (Table 3–1). State and Federal reporting requirements also become issues of concern if contamination is found.
- *Schedule Requirements*: The driving factor for the client's due date should be determined. Typical drivers include: a closing date for a transaction, need to make year end financial allocations, or regulatory deadlines.

TABLE 3–1   States with Regulatory Requirements and/or Guidance for
Site Assessments

California	Illinois	Minnesota	Oregon
Connecticut[2]	Maryland	New Hampshire[*1]	Pennsylvania
Delaware	Massachusetts[*]	New Jersey[*2]	Washington
Florida[*1]	Michigan	New York	Wisconsin

[*]Have specific Phase II guidance
[1]Limited to petroleum contaminated sites
[2]Require site assessment prior to property transfer
("Negative declaration requirement")

- *Budget Limitations*: Budget limitations should be defined and discussed at the proposal (pre-award) stage. Budget limitations may include: a pre-established budget, money being held in escrow for disbursement as the investigation is completed, or internal budget restrictions due to the value of the property transaction or the financial condition of the client.

3.2.1.2 Background Information Review   Project objectives and a preliminary scope can be further refined after available data is reviewed. The first step in developing any scope of work is to review available information to determine what is already known or presumed about the site. In the case of a Phase II investigation much of the background research was completed and summarized during the Phase I investigation, which may have been completed by another company. Also, there may be other reports completed for different reasons or by different parties which shed light on site conditions and can help in the design of the Phase II scope of work. Information gaps may be identified during review of background information. In such cases, additional focused research becomes the first task of the Phase II investigation.

The following list of items to consider during review of background information is designed to assist the user in determining the quality of information and research already available or completed and identifying data gaps as well as significant findings.

- Determine if the Phase I is a thorough study (Refer to Chapter 2). Is a list of references provided? How long ago was the study performed? Have site conditions changed that may affect the conclusions and recommendations?
- Determine if the purpose and objectives of the sponsoring parties and end-users for previous work are consistent with the current project purpose and objectives. Use data collected by others with care; in these instances, some degree of data validation is critical.
- Identify potential contaminants of concern. This information will be needed later to select analytical parameters and methods, and media to be sampled (refer to Section 3.3).

- Identify potential on-site sources of contamination. This information will be needed to select sampling/investigation locations, the number of samples, and the media to be sampled (refer to Section 3.3).
- Identify what, if any, background considerations are necessary. Unless there is evidence of a direct release, there will be a need to establish background conditions at the site.
- Identify possible sensitive receptors (e.g., wetlands, surface water bodies, drinking water supplies, basements, residences, schools or playgrounds, etc.). This information will be used to select the sampling media and locations.
- Identify possible migration pathways (e.g., groundwater, surface water, soil, or air). This information is an outgrowth of the sensitive receptor analysis and is needed to select the media to be sampled and the sample locations.
- Identify possible (or potential) completed exposure pathways (i.e., routes that contaminants may take from source to receptors). A completed pathway will include a source (release of a substance), a contaminated medium (groundwater, etc.) an exposure point (well, faucet, etc.), and a receptor population (children, etc.). If any one of these elements are missing, a complete exposure pathway does not exist.

3.2.1.3 *Scope of Work Development*   The scope of work will evolve as a result of the initial contact between consultant and client and the data gathering and review by the consultant thereafter. A formal scope of work, proposal, or work plan is the written result of the previously outlined sections and is a valuable tool for laying a firm site investigation framework.

The final scope of work should be viewed as something the client must determine with the assistance of the consultant because it relates directly to the extent of risk which the client is willing to accept. The more comprehensive the Phase II investigation, the greater the reduction in risks. But it must be noted that it is impossible to determine that a site is entirely free of environmental hazards. Findings are limited by the investigations that are performed.

The scope of work should define: data quality objectives, investigations to be undertaken and investigative methods, media to be sampled, analytical parameters and analytical methods, preliminary sampling locations, and estimated number of samples. Allowances for contingencies should also be included to account for different or unexpected conditions that may be encountered in the field. It should also include sections addressing the proposed schedule for completion, estimated project costs, and a list of assumptions used to develop the scope, schedule, and budget. Additional guidance for development of a suitable scope of work is included in Section 3.3.

The scope of work can take many forms: a formal proposal (ranging from brief, bare bones to fairly detailed), a work plan (very detailed, usually including detailed procedures to be employed, as well as Quality Assurance/Quality Control (QA/QC) requirements), or verbal agreements (developed as one proceeds; rarely done because of the potential for increased liability due to miscommunication or misunderstandings).

3.2.1.4 Data Quality Objectives Data Quality Objectives (DQOs) are scoping and planning tools applicable to every environmental sample collection effort and are a necessary step in the generation of a work plan. DQOs are quantitative and qualitative descriptions of data collected typically for one of the following purposes:

- identification of potential contamination,
- confirmation of contamination,
- characterization of contamination,
- evaluation of remedial alternatives,
- design of remedial alternatives,
- remedial actions,
- monitoring to comply with permits and regulations, and
- determination of background concentration levels.

As target values for data quality, DQOs are not necessarily criteria for acceptance or rejection of data. The DQO process generates a logical set of decisions that determines whether collection of samples is necessary. These decisions include types of samples to collect, including quality control samples, design of the sample collection effort including location and number of samples, analytical requirements, including precision, accuracy, comparability, completeness, and sensitivity of the method, and overall confidence level that the data will fulfill project needs. Decisions should fill existing data gaps and depend on intended data use. All steps of the process should be completed in sequence and be appropriately documented.

3.2.2 Project Mobilization

After the contract has been signed, the following list of items should be considered before actual mobilization for the field program. If not completed in a timely fashion (i.e., in advance), these items could jeopardize completion of a project on time and in compliance with the terms of contractual obligations.

- *Preliminary Report Outline Development*: A preliminary outline of the project final report helps to focus and organize the study early in the project and also allows the client and project team to better visualize exactly what to expect and/or accomplish. Development of a preliminary outline is key to focusing the project. The outline can be fleshed out and fine tuned as information is obtained once the investigation is underway.

- *Health and Safety Plan Development*: Hazardous waste site investigations, particularly those involving intrusive investigation, pose a multitude of health and safety risks to on-site workers and potentially to the surrounding community, including but not limited to: chemical exposure, fire and explosion, oxygen deficiency, ionizing radiation, biologic hazards, physical safety hazards, electrical hazards, heat stress, cold exposure, and noise. Health and safety plans anticipate such hazards and outline precautionary procedures to be followed to protect worker health and safety. Other related considerations include: OSHA training requirements, budgetary and schedule impacts of special Health and Safety procedures, and special equipment needs.
- *Site Access Issues*: Site access relates to both physical and legal access. Utility clearance both above and below ground for intrusive investigation points (call 1-800-DIG-SAFE and obtain as-built utility plans for the site) and physical access for equipment must be considered. Access agreements for property not owned by the client may need to be put into place (regulatory rights to access can be invoked in some instances).
- *Equipment Staging*: Equipment to be used needs to be identified based on activities to be carried out and what is available during the investigation timeframes. Related considerations include what utilities will be required and how they will be accessed.
- *Special Permit or Notification Requirements*: These may include permits to break pavement or investigate/excavate in the street or other public right-of-way or easement, wetlands permits, special notifications which might be required (for instance, EPA notification before excavation at a Superfund site; likewise, the local fire department must be notified of any tank removals and local officials should be notified if personal protective equipment will be worn during the investigation), and requests for a police detail (often required for work in or near roads). Prudent community relations may also require that notification of activities be provided to residents to forestall the "moonsuit" syndrome.
- *Active Facility Considerations*: Requirements may include security clearances (could take time), vehicle registration, special safety training requirements, and/or special work schedule (to prevent disruption of the normal facility workday). Many larger clients require drug testing for performance of field work.
- *Weather Considerations*: These impact health and safety issues (heat stress and cold exposure), require advance planning and preparation, and could affect the ability to implement specific tasks planned for the field program (field screening is adversely impacted in wet or humid field conditions and winter ice and snow can severely impact drilling and soil sample collection operations).
- *Special Transportation Requirements*: Schedules (and weight limits) for boats, barges, and ferry transportation should be obtained ahead of

time if necessary; long distance travel and equipment transport will require advance planning as will methods for transport of samples from the field to the laboratory.

- *Personnel Availability*: Phase II investigations involve the services of multiple disciplines. Coordination is required with department heads, office managers, and other project teams to assure personnel availability as required for the project tasks and schedule.
- *Subcontractor/Laboratory Availability*: The availability of subcontractors should be verified before the project schedule is accepted. The contract may be written to allow for delays due to subcontractor availability. Laboratories should be notified ahead of time of an upcoming shipment of samples to avoid exceeding analytical holding times because their workload was backed up with other samples.
- *Investigation-Derived Wastes*: A plan for characterization and disposal of any soil cuttings, decontamination fluids, or purge water that are generated must be developed. This may require special coordination with the client as well as drilling and/or waste removal contractors.
- *Emergency Response and Special Reporting*: The health and safety plan should contain contingency plans for dealing with on-site emergencies. The client should be kept abreast of the investigation progress and findings so that specific regulatory or insurance reporting requirements, if applicable, can be met in a timely fashion.
- *Community Relations*: Clients with particularly sensitive sites may want to put a community relations plan in place. In some cases, particularly those with state or federal agency involvement, community relations programs are a required part of the investigation process. Project personnel should be briefed on how to deal with questions from the public or media.
- *Team Briefing*: Everyone involved in the investigation should be familiar with the requirements and standard operating procedures, including QA/QC requirements, for the field investigation program.

3.3 FIELD PROGRAM IMPLEMENTATION

The data collection stage of the Phase II investigation represents the heart of the study. Data collection usually focuses on obtaining quantitative or semi-quantitative data from chemical analyses of selected media. Qualitative chemical or physical data such as groundwater pumping test results, soil grain size analyses, and dye tracer tests are often equally important. A detailed discussion of these qualitative methods is beyond the scope of this manual. Section 3.3.2 includes a discussion of a variety of intrusive and non-intrusive investigation techniques, sample collection techniques, and analytical methods. Although the reliability of the findings and conclusions of an investigation often increases with collection of more, and possibly better quality data, this does not mean that every scope of work should include everything mentioned in Section 3.3.2. Investigative/sampling techniques and analytical methods must be selected based on site conditions, study objectives, and budget limitations. The discussion that follows should help in determining what level of effort is appropriate for a specific investigation.

Data collection can be accomplished through invasive or non-invasive methods or a combination of both. Geophysical surveys (non-intrusive), for instance, can be implemented solely or in combination with invasive activities such as borings or test pits. When implemented alone, non-invasive results are more likely to be qualitative and inconclusive. Soil gas investigations are low invasive and require a lower level of effort than boring/well installation. Soil gas investigations, however, may not be conclusive in identifying sources of volatile compound contamination because the field analytical method is generally semi-quantitative and the sampled soil gas is only secondarily contaminated (volatile gases released from contaminated soil or the water table move through the soil column and are trapped in void spaces between soil particles). Soil boring and well installation with the collection of samples for laboratory analysis provide quantitative data, but also present a high degree of site invasion.

Design of a field investigation program begins with selection of investigative and sampling methods. This is determined in part by the level of confidence required for the investigation and is also dependent upon the objectives of the study. The field program planning process begins well before the project team gets in the field. The basic program framework is determined as the study objectives are laid out during early discussions with the client and review of background information. The proposed investigation framework is probably first developed in the initial scope of work. Further modification or refinement is entirely project specific, and in fact, may not be formally changed for a basic due diligence Phase II investigation. Larger scale regulatory-driven Phase II investigations often rework and refine the initially proposed scope through fee negotiation and work plan/sampling plan development. These produce a formal statement of the agreed upon scope that must be modified throughout the project if there are deviations required during the work.

The primary considerations for the development of a scope of work were discussed in Section 1. That section did not, however, detail the thought process involved in development of a sampling plan that will be the main section of the scope of work. Because the selection of analytical methods and sample location, type, and media is also relevant to implementation of the field investigation, discussion of the selection criteria is provided in detail in this section. Basic analytical methods and sample media and location selection should have been completed during the pre-investigation planning stage of the Phase II investigation. At the data collection/field investigation state, these selections are usually just refined and/or adapted based on actual field condition.

3.3.1 Sampling Plan Refinement

It is important to remember that the sampling plan should be viewed as a living document. Modifications can and should occur at any stage in the investigation process because of revised project or client objectives, new data, or changing regulatory requirements. Just as important, these changes should be documented to provide a complete record in the final report.

The focus of the field program should be on the collection of appropriate, reliable, and acceptable data. Environmental samples form the basis of the data collection effort. This effort should be designed with an eye toward the overall project objectives. Sampling programs can be broken down into three types: investigatory, delineative, or confirmatory.

Investigatory sampling programs are used to confirm potential areas of environmental concern (usually identified in the Phase I investigation or early in the Phase II process) and to reduce the number of contaminants of concern for these areas for future investigation activities. These programs generally include a limited number of samples from each area of concern but a longer list of analytes for each sample.

Moreover, the sampling data collected is sometimes used and needed for public health assessments. Most of the environmental information needed for these assessments is the same as that needed as part of an engineering site investigation plan. The U.S. Department of Health and Human Services has developed several guidance documents detailing their requirements (see References). The additional information generally needed is in the following categories:

- contaminant concentrations in all off-site media to which the public may be exposed,
- an appropriate detection limit and level of quality assurance/quality control in samples to ensure that the resulting data are adequate for assessing possible human exposures,
- discrete samples that reflect the potential range of exposure of the public,
- surface soil and sediment samples not deeper than 3 inches,
- more extensive biota studies and analyses of edible portions only,
- more ambient and indoor air sampling, and
- lists of physical hazards and barriers to site access.

Delineative sampling programs are used to define the areal extent of confirmed soil and/or groundwater contamination. This type of program generally involves a greater number of samples, but the list of analytes is limited to selected target analytes, such as PCBs for a transformer release characterization. Delineative sampling is often conducted as a precursor to remedial action. Traditionally, this type of program will involve the completion of numerous soil borings or test pits, followed by the installation of monitoring wells. Use of field screening technologies, both sampling and analytical, using temporary sampling ports and real time on-site analytical techniques are now frequently employed as a time- and cost-effective alternative. Examples of these techniques are soil gas sampling surveys, GeoProbe or HydroPunch type sampling techniques, and field screening or field analytical methods for volatile organic compounds (VOCs), total petroleum hydrocarbon (TPH) and PCBs. Virtually any analytical parameter can be tested for in a field laboratory equipped with the appropriate instrumentation.

The third type of program is confirmatory sampling that is used in conjunction with remedial actions such as tank removals, soil excavation, or free product removal from the water table. These sampling programs are usually driven by local, state, or federal regulations and include a specific number of samples submitted for a select list of target analytes.

The basic premise for any sampling plan is to determine how to achieve project goals and requirements using the lowest possible level of effort (i.e., at the least cost to provide the required amount of data). The answers to several key questions form the basis of any sampling plan.

- What kind of data is required?
- Where should the data come from?
- How much data is "enough?"
- Who will be using the data?
- What level of confidence in the data is required?
- What methods are available to acquire the data?
- What are the quality control objectives?

It is important to remember that the answers to these questions will be different for every investigation, and may, in fact, change over the life of a project. For example, review of existing data for a site may suggest that there is one underground tank located somewhere on the property. A non-intrusive geophysical survey is used as a first step to pinpoint the location of the tank. However, the results of the geophysical investigation may indicate that a tank farm of six tanks is present. Instead of proceeding directly to the proposed second phase of the investigation [a soil boring program with confirmation samples submitted for volatile organic compounds (VOCs) analysis], the scope of work and sampling plan is modified and a delineative soil gas sampling program is implemented to attempt to identify the composition and magnitude of tank contents that may have been released from the tank farm. This approach is determined to be more cost-effective than simply increasing the number of proposed borings and expanding the target analyte lists; and as long as the appropriate quality control measures are employed, the soil gas results are likely to be acceptable to the regulatory agencies.

The remainder of this section is an overview discussion of the primary field investigation techniques, sampling and sample selection methods, and laboratory analytical techniques commonly employed in Phase II field investigations.

3.3.2 Investigative Methods/Sample Selection Criteria

A wide variety of field investigation techniques may be used to accomplish the objectives of a Phase II investigation. Commonly used methods and the kinds of data they provide are summarized in Table 3–2. It is important to remember that no single investigative technique is more useful or more reliable or more effective than another when applied to the appropriate situation. Each method has its inherent strengths and weaknesses, which makes it applicable to some situations and not others. Selection of the appropriate technique should be performed by a professional familiar with the site geological characteristics, the goals of the sampling, and the investigative techniques available for use at the site.

A well-thought out project planning activity weighs the merits of the available technologies and results in a field investigation and sampling program that frequently uses several techniques, often in a phased approach, to maximize the value of the individual methods. There are a wide variety of published guidance documents and method papers that describe in detail the "how to" for each of these methods. The following sections briefly outline several key investigative techniques along with their applications and limitations.

3.3.2.1 Geophysical Studies Geophysical studies are often employed as a non-invasive preliminary screening method before performing invasive site investigation methods; or they may be implemented solely or in combination with invasive investigation techniques such as borings or test pits. The key to effective use of geophysical data is to understand the specific strengths and limitations for each method and to remember that the results of any geophysical survey are, by nature, subjective. Geophysical methods should be viewed as reconnaissance tools to guide

TABLE 3–2   Common Phase II Investigation Methods

Common Method	Information Obtained
Test Pits	• Subsurface soil conditions
	• Depth to groundwater
	• Collect soil/waste/groundwater samples
	• Soil permeability data
Soil Borings, Monitoring Well Installation	• All of above
	• Direction of groundwater flow
Aquifer Testing (Pump Test)	• Characteristics of aquifer
Groundwater/Surface water/ Sediments/Soil/Waste Sampling	• Presence, concentration, extent of contamination
Geophysical Surveys	• Geologic setting, soil, approximate depth to rock
Soil Gas Surveys	• Prediction of extent of VOC contamination in groundwater
	• Concentration of VOCs in vapor phase in soil

TABLE 3–3 Characteristics of Geophysical Methods

Method	Response Characteristic	Mode of Measurement	Depth of Penetration	Resolution	Raw Data Format
Ground Penetrating Radar (GPR)	Complex Dielectric Constant of soil, rock, pore fluids, and man-made objects	Continuous Profile .4 kM/hr. detail, 8 kM/hr reconnaissance (ground contact not necessary)	One to ten meters typical-highly site specific. Limited by fluids and soils with high electrical conductivity and by fine grain materials	Greatest of all six geophysical methods	Picture-like graphic display. Analog tape, digital tape
Electromagnetics (EM)	Bulk electric conductivity of soil, rock and pore fluids (pore fluids tend to dominate)	Continuous profiles to .5 to 15 m depth. Station measurements to 15 to 60 m depth. Some sounding capability (Ground contact not necessary)	Depth controlled by system coil spacing .5 to 60 m typical	Excellent lateral resolution. Vertical resolution of 2 layers. Thin layers may not be detected.	Numerical values of conductivity from station measurements. Stripchart and/or magnetic recorded data yields continuous profiling.
Resistivity Sounding (RES)	Bulk electrical resistivity of soil, rock and pore fluids (pore fluids tend to dominate)	Station measurements for profiling or sounding (Must have ground contact)	Depth controlled by electrode' spacing. Limited by space available for array. Instrument power and sensitivity become important at greater depth.	Good vertical resolution of 3 to 4 layers. Thin layers may not be detected.	Numeric values of voltage current and dimensions of array. Can plot profile or sounding curves from raw data.

72

Seismic Refraction	Seismic velocity of soil or rock which is related to density and elastic properties.	Station measurement (Must have ground contact)	Depth limited by array[1] length and energy source	Good vertical resolution of 3 to 4 layers. Seismic velocity must increase with depth—thin layers may not be detected.	Numeric values of time and distance. Can plot T/D graph from raw data
Magnetometer (MAG)	Magnetic susceptibility of ferrous metals	Continuous total field or gradient measurements. Many instruments are limited to station measurements. (Ground contact not necessary)	Single 55 gal drum up to[2] 6 m. Massive piles of 55 gal drums up to 20 m.	Good ability to locate targets.	Non-quantitative response from audio/visual indicators. Quantitative instruments provide motor or digital display (may record data)

1. Depth is also related to equipment capability.
2. Depth is very dependent on instrument used.

investigations and to augment data derived from direct investigation techniques.

Despite these caveats, geophysical studies can be a time-saving, cost effective method for providing qualitative subsurface information for a site. For example, they can be used for screening large areas for potential buried wastes, for focusing resources for intrusive investigation activities on anomalous areas, and for identifying or confirming the presence and configuration of underground storage tanks. Because they are generally non-intrusive, they are less likely to require the use of upgraded personal protective equipment during implementation and are generally less costly than invasive technologies.

Most of the methods employed in environmental site investigations were originally developed and/or refined by the petroleum exploration industry. Surface based investigation methods include electromagnetic techniques, electrical resistivity, ground penetrating radar, magnetic, and seismic refraction. Borehole methods (i.e., requiring an open borehole such as a monitoring well or a stable soil boring) include logging electrical resistivity (E logs) or radioactivity (gamma logs), or using a downhole video method to visually log the hole. Tables 3–3, 3–4, and 3–5 summarize the characteristics, applications, and limitations for each method.

3.3.2.2 Soil Gas Surveys   Soil gas surveys are gaining popularity as an effective screening technique for mapping the extent of VOCs, particularly low molecular weight halogenated compounds (solvents). These compounds can migrate (or partition) out of contaminated soil or groundwater and into the natural voids between soil particles. The natural tendency is for these gases to follow the path of least resistance and diffuse directly upward, and to a lesser extent, laterally, from the source of the VOCs. Thus, the soil gas concentrations should theoretically increase towards the source of the contamination or the area of highest concentration in those media. As such, soil gas contaminant mapping may be used as an indirect indication of the location of contaminant sources or the extent of soil or groundwater contaminant migration. The following are some of the advantages and disadvantages of soil gas investigations.

Advantages

- Generates extent of contamination data in a relatively short time.
- Low cost per sample location.
- Minimal disturbance to landscape.
- Minimal decontamination required.
- No investigation-derived waste (IDW) generated.
- Very adaptable to specific site conditions.

Disadvantages

- Detection limits may be too high.
- Specific compounds may not be readable by instrument.
- Field results are semi-quantitative.

TABLE 3.4   Typical Applications of Geophysical Methods

Application	Radar	EM	RES	Seismic	MAG
Natural Conditions:					
Layer thickness and depth of soil and rock	1	2	1	1	NA**
Mapping lateral anomaly locations	1	1	1	1	NA**
Determining vertical anomaly depths	1	2	1	1	NA
Very high resolution of lateral or vertical anomalous conditions	1	1	2	2	NA
Depth to water table	2	2	1	1	NA
Sub-Surface Contamination Leachates/Plumes:					
Existence of contaminant (Reconnaissance Surveys)	2*	1	1	NA	NA
Mapping contaminant boundaries	2*	1	1	NA	NA
Determining vertical extent of contaminant	2*	2	1	NA	NA
Quantify magnitude of contaminants	NA	1	1	NA	NA
Determine flow direction	2*	1	1	NA	NA
Flow rate using 2 measurements at different times	NA	1	1	NA	NA
Detection of organics floating on water table	2*	2*	2*	NA	NA
Detection & mapping of contaminants within unsaturated zone	2	1	1	NA	NA
Location and Boundaries of Buried Wastes:					
Bulk wastes	1	1	1	2	NA
Non-metallic containers	1	1	1	2	NA
Metallic containers					
—ferrous	2	1	NA	NA	1
—non-ferrous	2	1	NA	NA	NA
Depth of burial	2	2	1	2	2*
Utilities:					
Location of pipes, cables, tanks	1	1	NA	2	1
Identification of permeable pathways due to utility trench fill	1	1	NA	2	1
Abandoned well casings	NA	NA	NA	NA	1
Safety:					
Pre-drilling site clearance to avoid drums, etc.	1	1	2	NA	1

1—Denotes primary use
2—Denotes possible applications, secondary use. In some cases this may be the only effective approach due to circumstances.
NA—Not applicable
*—Limited applications
**—Not applicable in the context used in this document

There are several methods used to collect and analyze soil gas samples. All of these methods involve the placement of hollow, small diameter probes, driven manually, electrically, or hydraulically, to some optimum depth above the water table. The lower end of the probe is perforated to allow the gases to enter the probe. These probes may be temporary or permanent, depending on the needs of the project (e.g., preliminary site

TABLE 3.5 Susceptibility of Geophysical Methods to "Noise"

Source of Noise	Radar	EM	RES	Seismic	MAG
Buried pipes	1—will detect, but may affect data	1—only if close to pipe	1—only if survey is parallel and close by	2—only if survey is directly over	1—steel pipes only
Metal fences	NA	1—only if close to fence	2—only if survey line is parallel and close to fence	NA	1—steel fences only
Overhead wires (powerlines)	2—only if unshielded antennas are used	1	NA	NA	2—some mags respond
Ground vibrations	NA	NA	NA	1	NA
Airborne electromagnetic noise	NA	2	2	NA	1 to 2—(earths field changes)
Ground currents and voltage	NA	NA	2	NA	NA
Trees	2—only if unshielded antennas are used	NA	NA	2—(wind noise)	NA
Metal from buildings, vehicles, etc.	2—only if nearby & unshielded antennas are used	2—only if nearby	NA	2—only if nearby	2—only if nearby

Small metallic debris on surface or near surface (drums, drum covers, etc.)	2	NA	NA	NA	1—ferrous metal only
Large metallic debris on surface or near surface (drums, drum covers, etc.)	2	2	2	NA	1—ferrous metal only
Ground contact/electrode problems	2	NA	1	2	NA

Notes: 1—Very susceptible
2—Minor problem
NA—Not applicable

screening or long-term monitoring devices). A vacuum is applied to the probe and several pore volumes of gas are extracted to allow the probe to fill with ambient soil gas. A representative sample is extracted by inserting a glass syringe into the tubing below the vacuum source, by drawing the sample directly into a Tedlar sample bag, or by collecting the sample in a stainless steel adsorption tube. Syringe sample collection is appropriate for on-site direct injection gas chromatograph (GC) analytical methods. The Tedlar bag collection technique is also appropriate for field screening analyses and for off-site analytical laboratory analysis by GC or GC/MS methods. Use of adsorption tubes for sample collection is required for quantitative laboratory analysis of soil gas samples.

The results of a soil gas investigation are only as good as the method used to analyze the samples. Possible analytical variables include: the type of detector used with the GC apparatus, whether the samples were heated during the analytical process, the initial sample volume, and t .e target compounds selected.

3.3.2.3 *Invasive Soil Investigations*   Although soil gas investigations are appropriate for a number of situations, most Phase II assessments require invasive soil investigations for one or more purposes.

- Collection of soil samples or cores for lithologic logging.
- Collection of soil samples for laboratory testing.
- Characterization of lithologic and hydrogeologic parameters.
- Installation of monitoring wells or piezometer.

Development of an appropriate work plan for these types of investigations requires that a number of items be evaluated and addressed and that a number of decisions be made related to such things as:

- sample location selection–focused or grid,
- geologic and chemical characterization–continuous or interval sampling,
- drilling method–hollow-stem auger, rotary or others,
- alternative methods–cone penetrometer or test pits, and
- sampling method–split-barrel or thin wall tube or others.

Decisions are generally based on the goal of the investigation, cost and availability of equipment, suitability for the type of geologic materials at a site, and potential effects on sample integrity. The following section provides a broad overview of each of these factors and the benefits and constraints for each.

Sample Location   The intent of an invasive soil investigation is to determine if there is contamination present and to characterize and provide a generalized estimate of the limits of any identified soil contamination. Existing information and information obtained through non-invasive evaluations can be used to predict areas where contamination might occur. Where this information is available, sampling locations should be directed to those areas. Where there is no existing information, but an invasive investigation has been deemed necessary because of the strong possibility of contamination due to past operations at the site, a grid pattern of sampling locations is appropriate to evaluate a large area in a cost effective and efficient manner. Often a combination of specific siting (the length of the stream or channel bed) and gridding (samples collected only in the center of the channel and at a certain distance interval apart) will be the most appropriate method pattern of sampling, especially if contamination is believed to be due to spills or may have been spread by overland flow in drainage channels or streams.

Generally, for Phase II investigations, the purpose is to determine if there is contamination rather than identify the extent of contamination. Therefore, sampling is generally concentrated in areas suspected to be contaminated and the more focused pattern of sampling is appropriate.

Geologic and Chemical Characterization   While confirmation of contamination is one of the major goals of the Phase II investigation, the identification of the compounds causing the contamination and related breakdown products and geologic information about the site are also needed to develop an opinion of the extent of the problem and related health risks and remediation requirements. Geologic logging and chemical characterization of soil samples provide the most information about the type and extent of contamination.

Geologic logging of any test pit or bore hole gives an indication of the possible spread of the contamination. The more porous and loose the soils (sands, loams, etc.), the more easily the contaminants may migrate and spread from their source and the more potential for contamination of shallow groundwater. The more dense the soil (clays, shales, etc.), the slower the migration and spread of the contamination. Geologic logging can be done on a continuous basis or at intervals as the bore hole is advanced. Collection of soil cores is the preferred method for both logging and sampling since it is much more accurate than the sampling or logging of cuttings from drill methods that do not obtain cores. In general, core samples are collected at approximately 18 inch intervals. This provides a close approximation of continuous sampling while affording the most flexibility and efficiency of advancing the boring.

Selection of the chemical analyses to be performed is a function of the suspected contaminants. Close evaluation of existing data related to previous uses of the site, possible or probable materials spills, chemical breakdown products, and observed contamination should provide guidance on the most likely contaminants to be found. These are the contaminants of concern and samples should be screened for them. In addition, a number of samples collected from areas most likely to be contaminated should be screened for the broad range of hazardous compounds generally tested for at hazardous waste sites. Table 3–6 provides a list of these

TABLE 3–6   SW-846 Method 8240 for Volatile Organic Compounds

Compound	CAS Number[1]	Compound	CAS Number[1]
Acetone	67-64-1	cis-1,3-Dichloropropene	10061-01-5
Benzene	71-43-2	trans-1,3 Dichloropropene	10061-02-6
Bromodichloromethane	75-27-4	Ethylbenzene	100-41-4
Bromoform	75-25-2	2-Hexanone	591-78-6
Bromomethane	74-83-9	Methylene chloride	75-09-2
2-Butanone	78-93-3	4-Methyl-2-pentanone	108-10-1
Carbon disulfide	75-15-0	Styrene	100-42-5
Carbon tetrachloride	56-23-5	1,1,2,2-Tetrachloroethane	79-34-5
Chlorobenzene	108-90-7	Tetrachloroethene	127-18-4
Chlorodibromomethane	124-48-1	Toluene	108-88-3
Chloroethane	75-00-3	1,1,1-Trichloroethane	71-55-6
Chloroform	67-66-3	1,1,2-Trichloroethane	79-00-5
Chloromethane	74-87-3	Trichloroethene	79-01-6
1,1-Dichloroethane	75-34-3	Vinyl acetate	108-05-4
1,2-Dichloroethane	107-06-2	Vinyl chloride	75-01-4
1,1-Dichloroethene	75-35-4	Xylene (total)	1330-20-7
trans-1,2-Dichloroethene	156-60-5	1,2,3-Trichloropropane	NA
1,2-Dichloropropane	78-87-5	1,2,4,5-Tetrachlorobenzene	NA

[1]CAS Number—Chemical Abstract Services Registry number.
NA—Not available or not applicable.

TABLE 3–6 (cont.)   SW-846 Method 8270 for Semivolatile Organic Compounds by Gas Chromatography Mass Spectrometry (GC/MS): Capillary Column Method

Compound	CAS Number[1]	Compound	CAS Number[1]
Acenaphthene	83-32-9	2,4-Dinitrophenol	51-28-5
Acenaphthylene	208-96-8	2,4-Dinitrotoluene	121-14-2
Anthracene	120-12-7	2,6-Dinitrotoluene	606-20-2
Benzoic acid	65-85-0	Di-n-octyl phthalate	177-84-0
Benz(a)anthracene	56-55-3	Fluoranthene	206-44-0
Benzo(b)fluoranthene	205-99-2	Fluorene	86-73-7
Benzo(k)fluoranthene	207-08-9	Hexachlorobenzene	118-74-1
Benzo(g,h,i)perylene	191-24-2	Hexachlorobutadiene	87-68-3
Benzo(a)pyrene	50-32-8	Hexachlorocyclopentadiene	77-47-4
Benzyl alcohol	10-51-6	Hexachloroethane	67-72-1
Bis(2-chloroethoxy) methane	111-91-1	Indeno(1,2,3-cd)pyrene	193-39-5
Bis(2-chloroethyl) ether	111-44-4	Isophorone	78-59-1
Bis(2-chloroisopropyl) ether	108-60-1	2-Methylnaphthalene	91-57-6
Bis(2-ethylhexyl)phthalate	117-81-7	2-Methylphenol	95-48-7
4-Bromophenyl phenyl ether	101-55-3	3,4-Methylphenol	108-39-4
Butyl benzyl phthalate	85-68-7	4-Methylphenol	106-44-5
4-Chloroaniline	106-47-8	Naphthalene	91-20-3
4-Chloro-3-methylphenol	59-50-7	2-Nitroaniline	88-74-4

Compound	CAS Number[1]	Compound	CAS Number[1]
2-Chloronaphthalene	91-58-7	3-Nitroaniline	99-09-2
2-Chlorophenol	95-57-8	4-Nitroaniline	100-01-6
4-Chlorophenyl phenyl ether	7005-72-3	Nitrobenzene	98-95-3
Chrysene	218-01-9	2-Nitrophenol	88-75-5
Dibenz(a,h)anthracene	53-70-3	4-Nitrophenol	100-02-7
Dibenzofuran	132-64-9	N-Nitrosodimethylamine	62-75-9
Di-n-butyl phthalate	84-74-2	N-Nitrosodiphenylamine	86-30-6
1,2-Dichlorobenzene	95-50-1	N-Nitrosodi-n-propylamine	621-64-7
1,3-Dichlorobenzene	541-73-1	Pentachlorophenol	87-86-5
1,4-Dichlorobenzene	106-46-7	Phenanthrene	85-01-8
3,3'-Dichlorobenzidine	91-94-1	Phenol	108-95-2
2,4-Dichlorophenol	120-83-2	Pyrene	129-00-0
Diethyl phthalate	84-66-2	Pyridine	110-86-1
2,4-Dimethylphenol	105-67-9	1,2,4-Trichlorobenzene	120-82-1
Dimethyl phthalate	131-11-3	2,4,5-Trichlorophenol	95-95-4
4,6-Dinitro-2methylphenol	534-52-1	2,4,6-Trichlorophenol	88-06-2

[1]CAS Number—Chemical Abstract Services Registry number.

TABLE 3–6 (cont.)   SW-846 Method 6010 Inductively Coupled Plasma—Atomic Emission Spectroscopy Method Performance Data

Analyte	Analyte	Analyte	Analyte
Aluminum	Calcium	Magnesium	Silicon
Antimony	Chromium	Manganese	Silver
Arsenic (7060)	Cobalt	Mercury (7471 or 7470)	Thallium (7841)
Barium	Copper	Nickel	Vanadium
Beryllium	Iron	Potassium	Zinc
Cadmium	Lead (7421)	Selenium (7740)	

TABLE 3–6 (cont.)   SW-846 Method 8080 for Organochlorine Pesticides and Polychlorinated Biphenyls by Gas Chromatography

Compound	CAS Number[1]	Compound	CAS Number[1]
Aldrin	309-00-2	Endrin	72-20-8
α-BHC	319-84-6	Endrin aldehyde	7421-93-4
β-BHC	319-85-7	Heptachlor	76-44-8
δ-BHC	319-86-8	Heptachlor epoxide	1024-57-3
γ-BHC (Lindane)	58-89-9	4,4'-Methoxychlor	72-43-5
Chlordane (technical)	12789-03-6	Toxaphene	8001-35-2
4,4'-DDD	72-54-8	Aroclor-1016	12674-11-2
4,4'-DDE	72-55-9	Aroclor-1221	1104-28-2

Compound	CAS Number[1]	Compound	CAS Number[1]
4,4'-DDT	50-29-3	Aroclor-1232	11141-16-5
Dieldrin	60-57-1	Aroclor-1242	53469-21-9
Endosulfan I	959-98-8	Aroclor-1248	12672-29-6
Endosulfan II	33212-65-9	Aroclor-1254	11097-69-1
Endosulfan sulfate	1031-07-8	Aroclor-1260	11096-82-5

[1]CAS Number—Chemical Abstract Services Registry number.

TABLE 3–6 (cont.)   Method 8330 (Draft) for Nitroaromatics and Nitramines by High Performance Liquid Chromatography (HPLC)

Compound	CAS Number[1]	Compound	CAS Number[1]
Octahydro-1,3,5,7-tetranitro-1,3,5,7-tetrazocine (HMX)	2691-41-0	2,4,6-Trinitrotoluene (TNT)	118-96-7
Hexahydro-3,5,-trinitro-1,3,5-triazine (RDX)	121-82-4	2,4-Dinitrotoluene (2,4-DNT)	121-14-2
1,3,5-Trinitrobenzene (TNB)	99-35-4	2,6-Dinitrotoluene (2,6-DNT)	606-20-2
1,3-Dinitrobenzene (DNB)	99-65-0	o-Nitroluene (2-NT)	88-72-2
Methyl-2,4,6-trinitro-phenylnitramine (Tetryl)	497-45-8	m-Nitrotoluene (3-NT)	99-08-1
Nitrobenzene (NB)	98-95-3	p-Nitrotoluene (4-NT)	99-99-0

(1) CAS Number—Chemical Abstract Services Registry number.

TABLE 3–6 (cont.)   SW-846 Method 8150 for Herbicides

Analyte	CAS Number[1]	Analyte	CAS Number[1]
2,4-D	94-75-7	Dicamba	1918-00-9
2,4-DB	94-82-6	Dichlorprop	120-36-5
2,4,5-T	93-76-5	Dinoseb	88-85-7
2,4,5-TP	93-72-1	MCPA	94-74-6
Dalapon	75-99-0	MCPP	93-65-2

(1) CAS Number—Chemical Abstract Services Registry number.

contaminants and the general analytical method and category for each. This list is compiled from EPA SW–846 *Test Methods for Evaluating Solid Wastes, 3rd Edition,* which is the generally accepted reference for analysis of soils. The methods shown are those most commonly employed in performing these analyses. It is not intended to be inclusive of all possible analytical methods. For sites under EPA jurisdiction, analysis is generally controlled through the Contract Laboratory Program (CLP) which has its own list of parameters that is closely related to, but not exactly like, the SW–846 list.

Where metals contamination is suspected, it is also necessary to collect a representative background sample to use as comparison with samples collected from suspected contaminated areas. Because metals are naturally occurring and their concentrations vary substantially across the United States, comparison with local background is the best method for determining the presence and extent of metals contamination.

Drilling Methods A wide variety of drilling methods have been developed that could be suitable for Phase II assessments. Table 3–7 provides an overview of these methods and their applicability to various types of geologic formations. Selection of the appropriate drilling method should be based first and foremost on suitability for the type of geologic materials and the ability to collect appropriate samples. Most methods are especially effective for certain types of geologic conditions and each has its limitations related to types of samples that can be collected.

The hollow stem auger is the most commonly used method for borings in unconsolidated deposits (non-rock); air rotary is probably the most common for consolidated formations (rock). Selection of the appropriate drilling method should be carried out in cooperation with a qualified geotechnical consultant with input from local drillers who are generally knowledgeable about conditions that are most likely to be encountered at a site.

It is most important that every precaution be taken by the driller on site to prevent cross contamination from boring to boring or from a contaminated geologic layer to an uncontaminated one. This can result when equipment is not thoroughly cleaned between borings. When a boring is to be advanced through an aquifer, care must also be taken to assure that cross contamination of underlying strata and lower aquifers does not occur. The prevention of this is a factor of both care in cleaning and decontaminating equipment between borings and the installation of appropriate casing during the boring to seal off each strata as the boring is advanced.

Alternative Methods Two other methods are frequently used in the collection of both geologic and chemical data. These are test pits and cone penetrometer technology. Test pits are generally used in areas where buried materials are suspected. They are especially useful where there is the potential to bore through shallow areas of waste. A test pit is usually excavated using a backhoe or similar equipment. Samples are collected from the walls of the pit and from any uncovered wastes. Care must be taken during excavation to provide for the safety of the operation. The unexpected exposure of volatile wastes or rupture of containers of volatile materials can result in worker exposures to hazardous materials. In addition, there is a greater risk from flammable or explosive materials being uncovered during the excavation. Test pits are a fast method for

Table 3-7 Summary of Drilling Methods

Drilling Technique	Depth Limitation (ft)	Advantages	Disadvantages
Hand Auger	30	Mobility	Not useful in unconsolidated material below water table Not useful in cemented material Limited application in gravelly material Mixed samples
Power-Assisted Hand Auger Power Auger-Hollow Stem	80 300	Same as above Ease of soil sampling No fluids required	Same as above Not good in caving formations or those containing boulders Not useful when undisturbed soil samples are required
Power Auger-Solid Stem Power Auger-Bucket and Disk	100	Holes up to 3 ft. + in diameter Shallow holes above water table	
Cable Tool	1,000+	Low drilling fluid requirements Good definition of water-bearing zones Good in caving, high-gravel content material Good formation in samples	Slow Not good for small-diameter wells Must drive casing following bit
Mud Rotary	5,000+	Good cutting samples Can leave hole open during drilling Rapid drilling	Mud may plug permeable zones Not effective in boulder-rich sediments Not acceptable to EPA control of drilling fluids Lost circulation
Air Rotary	5,000+	Fast in consolidated formations No drilling liquids introduced into well	Small cuttings May be "watered out" in high-water zones Containment of drilling return difficult

tracking and identifying anomalies identified through geophysical testing. They also may present problems on-site because they result in large amounts of excavated materials and holes to be refilled.

Cone penetrometry involves the hydraulic pushing of a cone-shaped instrument into the soil and the evaluation of properties of the soil through electronic instrumentation. In addition, recent advances in the technology now allow for the collection of soil gas and groundwater samples using the cone penetrometer. It is best used for site characterization. Newer systems provide analytical equipment as a part of the system. This provides analytical information in the field which can be used to guide the placement of the next hole. The method works best in softer soils and at shallower depths. Continuous measurements of the geologic characteristics minimizes the potential for overlooking thin strata that could influence the migration of contaminants. Cone penetrometry is a cost effective alternative to boring in situations where one time testing is desired and where other conditions are appropriate for its use.

Sampling Method Criteria for selection of an appropriate soil sampling method are based on whether an undisturbed core is required, soil conditions (cohesion, stones, moisture), sample size desired, and depth of sample. These factors will determine what method is used for collection of the sample. Field instrument calibration, sample handling, preservation, storage, shipping, and analysis are discussed later in this chapter.

The most commonly used soil sample collection devices are the split and solid barrel. Split barrels (also known as split spoons) are tubes constructed of steel with a tongue and groove arrangement running the length of the tube, allowing it to be split in half. The two halves are held together by a threaded drive head assembly on the top and a shoe at the bottom. After the boring has been advanced to the proper depth, the barrel is fitted to the drill rod and driven into the formation for the specified depth (usually 18 inches). When the split spoon is brought to the surface, it is disassembled and the core can be evaluated and a sample collected. Generally geotechnical investigations sample an 18 inch interval for every five feet penetrated. Solid barrel samples are similar to split spoons, except they cannot be taken apart. A core extruder might be required to remove the core from the barrel. They are less widely used than the split spoon sampler. Another type of disturbed core sampler is the rotating core. The rotating core is most appropriate for collection of disturbed cores in dense, unconsolidated, and consolidated formations.

Sampling devices are also available for the collection of undisturbed samples where it is important to determine the properties of the materials as they occur in the ground. The thin wall sampler is frequently used for general soil sampling and is the best to use for continuous sampling using a five foot thin wall sampler placed down the stem of the auger. Gravel or cobbles in the soil can disturb the sample during collection or damage the walls of the sampler; therefore, it is best used for softer, unconsolidated soils without stones.

3.3.2.4 *Groundwater Monitoring Techniques*   When there is the potential that groundwater is contaminated, there are four basic questions that must be answered to provide adequate decision making information.

- How deep is the groundwater?
- Which direction is it flowing?
- Is there contamination?
- What is the level of contamination?

Monitoring wells are the primary means used for the collection of samples and other data required to answer these questions. They provide an access point into groundwater for the measurement of groundwater levels and the collection of samples for chemical analysis. To achieve these objectives, monitoring wells must be constructed to intersect the appropriate aquifer, cause minimal disturbance to the formation, be constructed of appropriate materials and in such a way as to prevent contamination of the groundwater sample due to inadequate sealing or leaching of chemicals from the well materials.

In addition to appropriate construction details, the monitoring well must be designed in concert with the overall goals of the monitoring program. Key factors that must be considered include:

- intended purpose of the well,
- placement of the well to achieve accurate water levels and/or representative water quality samples,
- adequate well diameter to accommodate tools for well development, aquifer testing equipment, and sampling devices,
- sufficient number and spatial separation of wells to achieve adequate coverage of the site, and
- long range plans for incorporation of the well into an ongoing monitoring program.

3.3.3 Sample Collection Techniques

The greatest level of effort associated with a Phase II field investigation is the acquisition of representative samples for analysis by the selected methods. There are a number of methods that can be used for the collection of soil or water samples. A partial listing of these methods is contained in Table 3–8. Selection of the appropriate method should be based on such factors as:

- material being sampled,
- analysis to be performed, and
- information to be acquired.

3.3.3.1 *Sample Types*   There are two basic sample types: grab and composite. Because each sample is analyzed individually, analytical costs

TABLE 3–8   Summary of Sampling Methods

Media	Method	Description
Soil	Scoops, spoons, shovels	Hand-operated stainless steel instruments used to collect surface or shallow soil samples. Yields disturbed sample.
	Augers	Auger bit rotating into soil and soil retained on the auger tip is brought to the surface and used as sample. Yields disturbed sample.
	Tube sampler	Similar to auger except closed or open tube with cutting tip is used. Tube is pushed into soil, then brought to surface. Yields relatively undisturbed sample.
	Split and solid barrel	Power driven split spoon allows for collection of core that can be opened to yield relatively undisturbed sample. Solid barrel yields undisturbed sample that must be extruded.
Groundwater	Open bailer	Hand or power winch operated, inexpensive. Bailer lowered into well, filled, then retrieved. Sample bottles filled at surface from bailer. Some loss of volatile compounds can result.
	Positive displacement pumps	Submersible pumps placed below static water level in well and pump sample to surface for collection.
	Suction lift pumps	Pumps placed at ground surface. Water drawn to surface by suction process. Some loss of volatile compounds can result.
	Sensors	Direct reading devices lowered to water level in well product instrument readout at surface.

should be considered when evaluating which method to use at a particular site.

A soil grab sample is collected from one specific location and depth and a water grab sample is collected from one location at one specific point in time. Grab samples are generally used when specific information related to specific locations or depths is required. They are the most commonly used method for the collection of groundwater samples and are also frequently used for soil samples.

Composite soil samples consist of a number of same size portions of soil collected from several locations and/or depths, all of which are mixed together and analyzed as one sample. Composite water samples have the potential for the added dimension of time when the sample is being collected from a well or from a flowing stream. Composite water samples are not commonly used for groundwater because the area of interest is generally so large that information obtained from composited samples is not detailed enough to allow characterization of the aquifer in a way to answer the questions previously defined as the driving purpose of the sample collection and analysis.

While soil, surface water, and groundwater are the most common samples collected, frequently other materials will also require evaluation and sampling.

- *Debris*—Discarded materials such as metal or fiberboard drums or containers are generally sampled by collecting a grab sample of materials inside the container.
- *Stream sediment*—Sediments in potentially contaminated streams are generally sampled to depths of 6 to 12 inches in areas where deposition of sediment is the most pronounced or most likely to have occurred or immediately downstream of the suspected or known contaminated water discharge point. Samples may be either grab or composite.
- *Stormwater*—Collection of stormwater is generally done by the grab sample method. Special effort should be made to collect the sample within the first 15 minutes of runoff since this is the time of highest contaminant concentration. Roof drains should be sampled in the same manner.
- *Leachate or seeps*—Grab samples are generally appropriate for evaluation of groundwater seepage or leachate since the characteristics of these sources do not generally vary over short periods of time.

In addition to the above, at some sites it may be necessary to take other types of samples. For instance, swipe samples are frequently collected for the determination of the presence of PCBs and bulk samples for the determination of the asbestos content of materials. In general, these additional types of samples are building and facilities related and become important when a facility is slated for demolition or renovation.

Swipe samples are collected from hard non-porous surfaces such as equipment, ducts, and floors. The samples are collected by wiping a surface with a gauze pad or piece of solvent soaked filter paper. The pad or filter is then placed in a clean container and shipped to the laboratory for analysis.

Bulk samples are the general method for collection of samples from materials that are potentially asbestos containing materials. In this method, a representative portion of the suspected material is collected by using a knife, cork borer, or other instrument to break loose a portion of the material. This portion is placed in a clean sample container and shipped to the laboratory for analysis. When a sample has been collected of material such as ceiling tile or piping insulation, the rough edge left on the remaining material must be sealed to prevent fiber release.

Samples may also be collected for the determination of lead content in paint. These samples are generally collected with a knife or other instrument that is used to pry loose paint chips which are then placed in a clean container and shipped to the laboratory for analysis. Alternately, on-site analysis may be performed on painted surfaces using X-ray fluorescent equipment.

3.3.3.2 *Quality Assurance/Quality Control*   The starting point for any sampling and analysis program should be the development of the quality assurance/quality control (QA/QC) plan. Sampling protocols are identified that will meet the data objectives of the project and specific procedures are defined to assure that the potential for errors is minimized at all stages of the sampling program. Major components of the QA/QC plan include the following.

- Documentation of control of samples through chain of custody forms and sign-offs.
- Details of decontamination procedures to minimize potential for cross contamination.
- Definition of sample collection procedures to assure comparability and representativeness of samples and reproducible results.
- Definition of collection, preservation, and transport procedures to minimize potential for contamination of samples.
- Details of duplicates, replicates, splits, and blank samples for determination of precision, accuracy, and reproducibility of results.

3.3.3.3 *Sample Handling*   Once a sample has been collected it must be preserved, packaged, stored, and shipped (using strict chain-of-custody procedures) in a way that assures the physical and chemical stability of the sample and its constituents. What is required varies depending on the type of media and the contaminants of concern in the sample. In general, samples should be placed in specially cleaned sample containers, preserved appropriately, sealed immediately, stored in a way that maintains the temperature of the sample at approximately 37° F, and shipped to the laboratory by overnight courier to assure that maximum holding times prior to analysis can be met. Chain-of-custody documentation should accompany the sample from collection through laboratory acceptance.

Each type of media and group of chemicals of concern has its own unique requirements for handling. Samples of both soil and water to be analyzed for volatile organic compounds must be placed in glass containers that are specially cleaned and the filled containers must not have any air space or air bubbles in them. Water samples for other types of organic compound analysis must be preserved with either acid or sodium hydroxide, depending on the compounds of concern, to assure the stability of the samples.

Allowable holding times prior to analysis vary depending on the compounds of concern and range from two to three days for certain specific compounds to up to 6 months for most heavy metals. The sampling plan for a specific site should be tailored to assure that collected samples will be handled using the appropriate containers, preservation, and storage methods and that analytical holding times are defined and agreed to by the laboratory completing the analysis.

3.3.4 Analytical Methods

Selection of the appropriate analytical methods to obtain the necessary information for a site assessment requires an understanding of the various levels of information that can be obtained. In general, the more sophisticated the analytical method, the lower the detection limits, the more accurate the results, and the more costly the analysis. The EPA has identified and approved, through its Contract Laboratory Program, analytical procedures for the analysis of all types of media for the compounds generally of concern. This information is contained in two federal publications, *40 Code of Federal Regulations (CFR) Part 136* and *Test Methods for Evaluating Solid Waste, 3rd Edition*, EPA/530/SW–846.

Once the parameters of concern have been identified, the selected laboratory can assist in the selection of the most appropriate analytical method and quality control analytical requirements.

3.3.4.1 Field Screening Once the compounds of concern have been identified for a site, the appropriateness of field screening of samples needs to be determined. Field screening techniques generally provide much less accurate and less sensitive analytical data than laboratory analysis. But they are extremely useful for gross determination of the presence and extent of contamination. Advantages of field screening include: results can be obtained almost immediately as opposed to days or weeks for laboratory analysis and the low cost per sample allows for further evaluation of more samples and/or reduction of overall analytical costs.

Field screening techniques range from simple direct read instruments such as the photo ionization detector which provides direct readout of total organics in air to sophisticated, mobile mass spectrometric gas chromatographs. In addition, numerous companies now offer simple field test kits for specific compounds such as PCBs and total petroleum hydrocarbons. The use of these methods at a site can greatly reduce the analytical costs since samples sent for laboratory analysis can be field screened to select only those which exhibit higher concentrations. Use of field screening during actual clean-up operations is generally for the opposite reason—to identify samples that are most likely to have contaminants below the clean-up level and determine the limits of the media to be treated. Field analysis alone cannot be relied upon to assure that a site is either clean or contaminated. Confirmation is required through laboratory analysis where strict quality control provides assurance that the results accurately reflect actual conditions and are not easily challenged should litigation arise. It is just as important to document that areas are clean as it is to document contamination. These field tests, supported by selective laboratory analysis, can be a cost effective procedure to provide this documentation.

3.3.4.2 *Laboratory Analysis*   Analysis of samples in a qualified laboratory with an established quality control program will provide the most assurance that actual conditions have been identified and results can withstand the tests of litigation that might arise. Selection of a laboratory is, therefore, of prime importance in the overall development of the site assessment investigation team. The first criteria should be the capability of the laboratory to perform the required analyses and their certification for such analyses in the state in which the investigation is being conducted.

In addition to state certification, the EPA has established its Contract Laboratory Program to address these needs for its investigation of Superfund sites. This program includes extensive quality tests for the laboratory and provides the highest level of documentation and quality assurance currently available. In most site assessment projects, this level is not warranted. But the selection of a laboratory should include an evaluation of their quality control program, as well as their capability and capacity to perform the required analyses. Discussions should be held with the laboratory to determine their method of data presentation, the availability of quality control data, and their current capacity to perform the required analyses.

Note that cost of analysis has not been addressed in this discussion. Costs will vary and, of course, will be a factor in the final selection of the laboratory; but unless the qualifications of the laboratory have been evaluated, low cost can be false savings and result in erroneous and erratic data that could jeopardize the integrity of the site assessment.

3.3.5 Data Interpretation

The challenge of the Phase II investigation is that it requires that the environmental specialist make critical decisions based on a limited set of data. Beyond the background information regarding site history and operations, general environmental setting available from public sources, and regulatory files, the site-specific data typical of a Phase II field program may be only a small number of chemical analyses of on-site media and groundwater elevations. However, maximum value can be derived from the limited data set if it is evaluated in light of other information, such as a receptor analysis, and if the limitations of any data interpretations are understood and consistently reported.

The aim of data evaluation is to address several major questions.

- Do the data confirm or disprove the presence of a release of hazardous material? If confirmed, what is the nature and m of the release, and is it from an on-site source or from off-erty? What media are impacted?

- If the data confirm the presence of oil or a hazardous material, what is the public health significance? Is there potential for human contact?
- If a release of oil or a hazardous material was present but undetected on-site, or if a release were to occur in the future, what level of environmental liability could be incurred? Are there migration mechanisms and pathways for oil or hazardous materials to migrate readily from an on-site release point to a sensitive receptor, such as a drinking water supply well or a sensitive wetland?
- Is further data collection warranted and cost effective? The answer to this question, of course, depends not only on an objective identification of any data gaps, but also the objectives and resources of the client commissioning the study.

The following discussion focuses not only on these questions but also on understanding the uncertainties inherent in interpretations and conclusions. Such uncertainties translate to a level of risk for the end user of the investigation and, therefore, must be clearly understood and conveyed.

3.3.5.1 Hydrogeologic Setting

Site Hydrogeologic Framework   An evaluation of the hydrogeologic setting should provide an understanding of the nature of soils and/or bedrock beneath the site. If there is a possibility that groundwater beneath the site has been or potentially could be impacted (e.g., if the depth of groundwater is "relatively" shallow), characterization of the shallow aquifer should also be a component of this assessment. A Phase II field program typically includes no more than three to five soil borings with the installation of shallow monitoring wells. Because the primary purpose of these borings and wells is to check for a release from a known or potential source, the location and depth may not be optimal for interpretation of the local hydrogeologic framework. Therefore, the data obtained from the field program (boring logs, depth to groundwater measurements, other field observations) should be augmented by available background information to arrive at a more complete picture of the site. For example, shallow soil borings may only extend to ten feet below grade, but the soils identified by the logs may be correlated with a regional stratigraphic formation. U.S. Geological Survey maps and Water Resources and Soil Conservation Reports may expand the understanding of a formation including its lateral extent, variability and thickness, the nature of the underlying formations, and the extent to which it is used as an aquifer.

Important considerations for this aspect of the data evaluation include the following.

- Is the site underlain by fill? What is the distribution of this material? Does it represent a potential source of imported contaminants onto the site?

- What is the nature of any unconsolidated materials beneath the site? Does it represent a ready migration pathway for contaminants released at or near the surface to enter the groundwater? Or, is the site underlain by a dense clay such that even if a release were to occur, it would be unlikely to impact an aquifer beneath this clay layer?
- What is the depth to groundwater beneath the site? Is groundwater so deep that if a release of oil or hazardous materials were to occur at or near the surface, the liquid would remain as dispersed, residual saturation within the unsaturated zone and not reach the water table?
- Is there a shallow aquifer beneath the site that is locally or regionally utilized as a groundwater source? Is the site over an aquifer that has been designated a sole source? Do published sources indicate that groundwater usage is limited to a deeper aquifer beneath the area, separated from the shallow groundwater system by an aquitard? What is the potential for vertical communication between the shallow and deeper systems? Is there a perched aquifer? How is it utilized?

Groundwater Flow   An understanding of groundwater flow conditions is key to understanding the potential impacts of an on-site release of oil or hazardous materials. An interpretation of the direction of groundwater flow can address, in a general way, two important questions.

- If a release to groundwater has occurred on the site or were to occur in the future, where would contaminants migrate and could on-site or off-site receptors be impacted?
- Is there a potential for a known or suspected groundwater contaminant plume derived from an off-site source to migrate onto the subject property?

To evaluate the direction of groundwater flow across the site, a groundwater contour map should be constructed. Generally, wells installed in a Phase II investigation are shallow, so this map would typically represent a water table map. The map is constructed from depth to groundwater measurements which are converted to elevation data. Such data may either be relative to mean sea level or another established regional benchmark, or they may be tied into an arbitrary benchmark on site. Keep in mind, however, that the water elevation data are only as good as the elevation survey used to calculate the elevations. Vertical survey data (as well as the depth to groundwater measurements) should generally be accurate to 0.01 inch. The accuracy of horizontal locations is not as critical, but locations are generally plotted to an accuracy of about one foot. Also, it is critical to utilize only water level measurements collected concurrently to prepare a groundwater contour map. Misleading contours may result from combining information from more than one measurement event. However, it may be possible, for the very shallow aquifers, in the absence of other information, to use topographic features to assist in the construction of groundwater contour maps. For example, surface water elevations from topographic maps (if surveyed to the same coordinate system) can be used in a very general way to approximate the direction of groundwater flow.

In many Phase II field programs no more than three monitoring wells have been installed, and in most cases these wells are screened in or across the water table. This permits straight line groundwater contours to be drawn. Only if other information is available (e.g., surface water elevations, topographic contours, presence of pumping wells) can the groundwater contour lines depart from being linear. Groundwater contour lines are generally not at a higher elevation than the topographic contours except where artesian conditions exist. Figures produced should include the date of the measurements. Contour lines should not be extended beyond the areal limits of the data although it may be acceptable to indicate contours as dashed lines beyond the data limits if there is additional information (such as topographic inferences) to substantiate these extrapolations.

Also note that monitoring wells installed may actually penetrate and sample a perched water horizon. One indicator of a perched water horizon is unexpected variations of groundwater elevations. Elevations of perched water cannot be meaningfully compared with the true saturated zone water table.

If floating product is detected in monitoring wells, field data collected should generally include depth to the product/air interface, as well as the depth to the product/water interface. It is important to note that the floating product displaces the water table and, therefore, the measured depth to water is not representative of the actual piezometric surface. An inferred piezometric elevation may be calculated if the thickness of product and its density (specific gravity) is known or can reasonably be estimated.

If the groundwater contour map is based on only a limited number of points, the groundwater gradient calculated as the slope of the piezometric surface (usually the water table) will represent the average horizontal hydraulic gradient across the mapped area. Vertical gradients are rarely considered during this phase, since deep monitoring wells generally are not installed at this stage of a site investigation. Estimated hydraulic gradients are generally not useful in a meaningful way to estimate the rate of groundwater flow of contaminant transport rates at the site unless aquifer properties of porosity and permeability have been reliably measured or estimated. These limitations can inject significant uncertainties in the calculation of travel times of plumes. The issue of contaminant transport is further compounded by retardation (i.e., contaminant transport at a rate slower than the groundwater flow rate). In some cases, the best that can be determined is the groundwater flow direction and velocity.

Once the direction of local groundwater flow is established or approximated, broader questions can be addressed: What is the source of on-site groundwater, and where is groundwater produced from potential on-site contaminant sources migrating? Is downgradient groundwater used for industrial purposes? Are there private or municipal water supplies downgradient from the site representing a serious source of potential liability should a release occur? Are there other sensitive receptors downgradient of potential on-site contaminant sources? If wetlands are near the site, what is the likelihood that potentially contaminated groundwater or contaminated surface runoff will discharge to these areas?

Stream Hydraulics Local rivers and streams may serve as hydraulic boundaries. In many cases, they are a discharge point for local groundwater. However, streams may be "losing" or "gaining" streams, and it may be misleading to always assume that they are discharge zones. It is important to consider the overall topography, as well as any engineered alterations. For example, a dam downstream could reverse the local hydraulic gradients on the site.

Tidal effects may be present in esturine areas, although there may be very little manifestation in water table fluctuations, except in areas immediately adjacent to the tidal water body. Nevertheless, there is a potential for river water to infiltrate onto the site. This would be enhanced if there is any pumpage of groundwater on-site.

3.3.5.2 Analytical Data Evaluation Interpreting analytical data may be the single most important activity conducted within a Phase II investigation. Since the usual goal of any Phase II is to determine if contaminants exist or not and at what risk level, improper evaluation and interpretation can place the environmental specialist and the client in potentially difficult and sometimes disastrous situations. While no manual can provide a fool-proof method for accomplishing this, the discussion below provides some common techniques to evaluate data.

Most importantly, regulations relevant to the study and to which data should be compared against must be identified. Many regulations or the agencies enforcing them will have standards and criteria for individual or compound classes and also guidelines, both written and unwritten, for interpreting data.

Some of the more important questions to be answered and findings to be made by evaluating and interpreting data are as follows.

- Were any contaminants detected?
- What contaminants were detected?
- What are background levels?
- Were any of the contaminants detected at concentrations above accepted background levels?
- Are standards and criteria available for those contaminants detected?
- Were the contaminants that were detected above background also above government standards or criteria?

- To what degree are contaminants above background or standards and criteria?
- For those contaminants above background or above standards and criteria, is there a reportable quantity criteria? Is it set by mass (weight) or by concentration detected?
- Are the detected concentrations a risk to human health or the environment?
- If a contaminant source is unknown, does the distribution of the contaminants indicate what and where the source is?

The evaluation and analysis of data are not restricted to these questions. The environmental specialist also needs to review the data and the answers to the above questions in the broader context of "are the contaminants mobile, soluble, exposed to surface contact and wind erosion, accessible, etc." and determine which characteristics of the contaminants will be crucial in determining whether there is a risk.

If the data are properly tabulated and evaluated, the concentrations detected can be put into the context of where they physically are located in the environment, how they will be transported and to where, and how it will affect human health or the environment.

Data Validation   Data validation is a process that is used to review results from chemical analysis to determine if the analyses have been performed within acceptable parameters. Data validation will ensure that the analytical results are in compliance with specified DQOs and adequate for their intended use. The specific criteria that are reviewed during data validation include:

- holding times,
- blank contamination,
- initial and continuing instrument calibration,
- surrogate recovery,
- duplicate precision,
- matrix spike recovery,
- laboratory control samples, and
- target compound identification.

The results from data validation are usually presented in three tables. The first table includes all data as reported by the analytical laboratory. These data include qualifiers that may have been assigned by the analytical laboratory. The second table presents data that have been validated and given EPA designated qualifiers. The third and final table presents all unqualified data.

Quality control criteria have been established by the EPA for approved EPA analytical methods. Some state agencies have additional criteria. Data validation is used to establish what data have or have not met quality control standards. Data are assigned qualifiers based upon results of a data validation review process. These qualifiers designate which analytical results can be used for required applications with a known level of confidence. All data are valid and acceptable except those that have been qualified. The data qualifiers that are assigned to data include the following.

J = Estimated concentration because all of the required quality control criteria were not met.

U = Not detected above Contract Required Detection Limit (CRDL) or Quantitation Limit (CRQL).

R = Results were rejected because of serious quality control deficiencies.

UJ = Quantitation limit was estimated because quality control criteria were not met.

JN = Presence of an analyte as tentatively identified and the associated results represent an approximate concentration.

Two things should be noted by all data users. First, the "R" qualifier flag means that the associated value is unusable. Due to significant quality control problems the analysis is invalid and provides no information as to whether the compound is present or not. Second, it is important to keep in mind that no compound concentration, even if it has passed all quality control tests, is guaranteed to be accurate. Strict quality control serves to increase confidence in data but any value potentially contains error.

The extent of data validation efforts depends upon the objectives of the investigation. For most Phase II investigations, a simple review of the laboratory data packages against the first four criteria noted above can be done by many environmental specialists. More detailed validation efforts should be conducted by staff specifically trained in data validation procedures as outlined by EPA or the respective state agency.

As previously noted, EPA has established a technical data system referred to as Data Quality Objectives or DQO's. Within this system, the environmental specialist evaluates the end use of the analytical data and determines the minimum degree or "level" of data validation required by the U.S. EPA. For Phase II's, the lowest level is commonly acceptable. The system also refers to the methods of analysis, which have been discussed earlier in the Data Collection Section.

Additionally, field collected data may require validation. The purpose of this is to ensure that sample collection, instrument calibration, and documentation have been done in accordance with DQO and QA/QC requirements.

Comparison to Background Comparing analytical data to background levels provides a valid method to potentially eliminate from concern various compounds detected at a site. In general, the effort to compare data to background levels is limited to inorganic species of compounds, specifically metals such as lead, mercury, arsenic, chromium, etc., all of which are naturally occurring in the environment. The key word here is "naturally occurring," as distinguishable from man-made but ubiquitous in the environment.

Although some of the inorganic compounds are natural, like lead, they also have been concentrated in the environment via human activities. Lead concentrations are commonly higher in soils near roads than in areas that do not have vehicular traffic as a result of exhaust emissions. Does that mean that the concentration of lead in soils near roadways is considered background for that kind of environment? Obviously, background can mean many things to many people. Common sense plays an important part in the interpretation and each site should be evaluated separately on a case-by-case basis. What was background for one site and, therefore, deemed acceptable for that compound, may not apply to another site.

The first step in evaluating data against a background is to determine what detected compounds/elements have background values established for them. This is first evaluated on a global scale (often referred to as the average soil concentration), then on a regional scale (the state the site is in), and lastly and preferably the local area of the site (county, city, or geomorphic region). Background data, however, is more readily available for the global or regional scale than for local areas. As more work is conducted and the need increases, data on the local level will likely increase.

Comparing data to background levels must be carried out on a sample basis as well as on an aggregate basis. Figure 3–2 depicts concentrations of lead detected in eight soil samples from a site, along with background values for lead. From the figure, it can be determined that most of the samples contain lead at concentrations below both the national and local averages for background. Only one sample (#SS008) shows any dramatic exceedance of the background levels. On average, site concentrations for the first seven samples would indicate that lead is well within background levels. This could lead to the opinion that lead, in general, is not a problem at this site.

However, the large value for lead in one sample is of concern. Concentrations of lead above 500 parts per million (ppm) are considered a health hazard for children, and other lead effects are well documented for other population groups. Thus, the concentration of lead in sample #SS008 poses a potential health risk. Does that mean the site is contaminated with lead and subject to governmental notification and site remediation? Further evaluation is necessary before that question can be answered. Where on the site was it collected? If the soil was collected from a location adjacent to a major roadway, it might be concluded that it is from vehicular exhaust. If collected from next to an old building, it might indicate that the soil around the building contains lead paint deposits and might warrant additional Figure 3–2 testing. If the roadway scenario is true, further testing or concern may not be warranted. Although the concentration is of a health concern, the site owner is not likely to be held responsible for roadway vehicular exhaust sources, only those sources

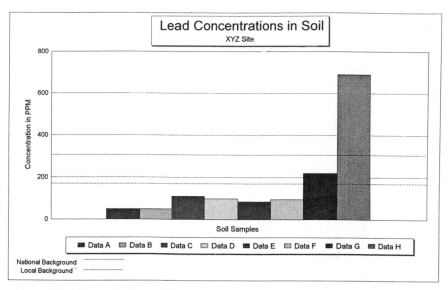

Figure 3.2—Example Data Plot

likely related to activity on the site for which the owner is responsible. If the source of high lead concentrations is unknown, then the site owner will likely be held responsible for any liability associated with it. That is not to say that one high lead value will represent a major environmental problem, it still may be possible to eliminate concern based on experience with similar situations, regulators of similar problems, or through a screening risk assessment where an average concentration can be used for exposure analysis.

Constructing a similar graph for every inorganic compound or element detected is not necessary for Phase II studies. The intent here is to draw the kind of mental image that the environmental specialist needs to create for comparing the data. Presenting a simple table (Table 3–9) that shows the average and high value for each compound/element and the corresponding background value(s), will convey a similar picture on one page.

Evaluation and Interpretation of Analytical Data The simplest method for evaluating analytical data is to compare those compounds detected or parameters measured to governmentally established standards, criteria, or guideline levels, or natural and site background levels as discussed above. However, most of the compounds that will be detected will be organic compounds not usually found "naturally" in the environment. Even the "ubiquitous in nature" approach will not likely suffice for most compounds because the concentrations of the local background may still be lower than detected on-site.

This leaves the investigator with evaluating obvious site-related contaminant concentrations as to whether or not there is a problem to report

TABLE 3-9 Example Summary of Analytical Data XYZ Site, Anywhere, USA

Compounds	Media	No. of Samples Analyzed	No. of Samples with Detectable Concentrations	Highest Concentration Detected (VOC – ppb) (Metals – ppm)	Average Concentration (VOC – ppb) (Metals – ppm)	Background Concentration or Standard	No. of Samples Exceeding Background or Standard
Volatile Organics						Standard	
Acetone	GW	1	0	0	0	250	0
Chloroform	GW	1	0	0	0	10	0
Toluene	GW	2	2	210	105	2	2
Benzene	GW	2	2	60	30	100	0
Xylene	GW	2	2	300	150	500	0
Inorganic Compounds						Background	
Arsenic	Soil	3	3	600	275	300	2
Lead	Soil	3	3	10	4	3	3
Mercury	Soil	3	3	6	6	73	0

GW—Groundwater
ppb—Parts per billion
ppm—Parts per million

or clean-up. A large number of the tested compounds will have established standards or criteria, but many will not. The EPA and other agencies are continuously developing and making available criteria such as the water related Maximum Contaminant Levels (MCLs) and Soil Screening Levels (SSL). For those compounds without standards, the environmental specialist should ascertain if the state or federal agencies have any guidelines or rule-of-thumb levels that they have used.

Prior assessment experience as to what the regulators have allowed in the past and whether a trend has occurred in the allowable limits (up or down) should be used when no guidelines or standards are available. These site by site precedents may provide valuable guidance on how to interpret current site data. When taking this approach, it is crucial that the client be informed of assumptions in this regard and the limitations of an opinion based on past experience rather than regulated standards and guidelines.

When faced with the need for additional basis for an opinion on unregulated compounds or ones without standards, etc., a risk assessment may be required. The risk assessment would estimate the likelihood for undue harm to human health based on a variety of exposure scenarios (i.e., children playing in the soil over a certain time frame).

In formulating and presenting an opinion, the analytical data should be placed in tabular form by media sampled, compounds/elements/parameters tested for and finally the comparison to established standards or criteria. Evaluation of data by the different media sampled is a relatively easy way to avoid confusion when dealing with various methods and detection limits of analyses. The data should be evaluated (and presented in the report) for each media:

- groundwater,
- surface water,
- surface soil,
- subsurface soil,
- air,
- sediment (sometimes combined with soil), and
- wastes, drums, & lagoon samples.

For each media, the analytical data would be evaluated for each group of compounds tested for parameters measured:

- field screening data (PID or FID),
- field parameters testing (i.e., conductivity, pH),
- inorganic compounds,
- volatile organics (VOCs),
- semi-volatile organics (SVOCs),
- PCBs & pesticides,
- oil/petroleum products (i.e., kerosene, diesel, etc.), and
- mixed and low level radiological wastes.

For any given Phase II, it is unlikely that all media will be sampled/tested for all compounds/parameters. The format of presentation in the report should be to first present the facts (the analytical data itself) in a format for interpretation (data by media and compound groups), then present findings/opinions (is it a release, hazard, etc.).

Data should be evaluated in the following ways.

- Of the samples analyzed, what number contained that specific compound?

    Benzene was detected in 5 of the 13 groundwater samples . . . .

- What was the concentration range detected?

    At concentration ranging from 15 to 230 parts per billion (ppb) . . . .

- If sufficient numbers of data are available and the statistics are meaningful, then what is the distribution of data?

    Benzene concentrations in the majority of samples (80%, 4 out of 5), were below 30 ppb, with a mean of 22 ppb.

    (This kind of evaluation will help determine the importance of the concentration, as previously shown with lead concentration versus background.)

- What standards, criteria, and/or guidelines apply?

    The benzene standard in groundwater for drinking water purposes is 5 ppb; all samples with benzene exceed this level.

- What is the spatial distribution of contaminants at the site, both in planar and vertical directions?

    Samples containing benzene were collected almost entirely from the north side of the site, adjacent to the loading dock at a depth of 2 feet or less.

This sets the stage for findings and opinions because the facts will have already been presented in a logical fashion, making opinions well founded. As can be seen, this kind of evaluation will also help the reader arrive at his own conclusions or formulate his own opinions.

In comparing data to any standard, etc., it is important to understand for what purpose the standard was set. For example, using a standard for drinking water for groundwater evaluation needs to be put into perspective. Groundwater in the site vicinity may not be used for drinking, or groundwater extraction points (i.e., drinking water wells) may be miles away. The concentration set for a given compound is for the concentration at the point of ingestion, i.e., the faucet. The sample collected from groundwater may be hundreds or even thousands of feet from a drinking water well or additional thousands of feet of distribution pipes and municipal treatment systems may be encountered before it reaches the faucet. To assume that the on-site sample contaminant concentration will remain the same until it reaches an ingestion point is unlikely and unreasonable. However, in the absence of more definitive standards, this may be the required evaluation criteria. Conversely, it may be a short distance to private wells with short distribution lines and possibly only particulate filters. In these cases, assuming the worst case of no reduction of concentration is in the best interest of everyone.

Many other types of standards or criteria are available for evaluating data in a variety of media and for a variety of purposes. Using all of these in concert will help in the selection of the best value for comparison. Where federal, state, and local standards are available, they are applicable. Often these standards contain widely varying criteria. In most, if not all cases, it will be necessary to evaluate site data related to the most stringent of these criteria.

3.3.5.3 Screening Level Risk Assessment   Once the nature and extent of contamination have been determined, the potential impacts to public health and the environment can be evaluated by a screening level risk assessment process. The objectives of screening level risk assessment are to identify public health and environmental receptors that could be exposed to contamination and to assess the magnitude of risks associated with the potential exposures. The EPA has developed a number of reference documents for use in performing all levels of risk assessment. Also, some states (e.g., Massachusetts) have also developed detailed methodologies to conduct risk assessments for sites located within those states.

Sensitive Receptor Analysis   The identification of potential receptors is a critical step in the risk assessment process. This identification can be achieved by surveying the site area to determine 1) those groups of people who, by their location and activities, could come in contact with contaminants at their source or in transport media, and 2) those environmental habitats that might receive contaminants and thereby have potential for exposure of plants and/or animals, especially threatened or endangered species. The identification of potential human receptors should consider activities associated with both current and future land uses.

The most commonly evaluated human receptors include residents, trespassers, on-site workers, and utility or construction workers. The resident receptor group is the most exposed receptor group, and, therefore, often serves as the focus of the screening assessment. Realistic site-specific exposure pathways would generally include incidental ingestion and/or dermal contact with soil and ingestion and/or dermal contact with groundwater used as drinking water. Depending on the facility, another realistic site-specific exposure pathway would be inhalation of contaminated air.

The magnitude of the health risks is generally screened by comparing concentrations in soil and groundwater to published risks related to soil and groundwater concentrations. These may include contaminant concentrations that are associated with both significant and insignificant health risks, as characterized by EPA or state regulatory agencies, EPA drinking water standards or Maximum Contaminant Levels (MCLs), and state regulatory agency drinking water standards or guidelines.

Screening Level Risk Assessment Use   The most appropriate use of published risk-related (sometimes called benchmark) residential scenario concentrations is in the evaluation of sites with current or potential residential use. However, even when the land and groundwater use is not residential, a comparison of site concentrations to these residential benchmark concentrations can be very useful. If the non-residential setting site concentrations are lower than the residential benchmark concentrations, then significant health risks at the site would probably not exist. If in the non-residential setting the residential benchmark concentrations are exceeded, a more detailed site-specific risk assessment may be in order. Some states have benchmark soils concentrations for industrial land use as well. These values can be used as the screen for industrial areas.

There are some site conditions that might be identified during the field program for which a screening level risk assessment approach may not be completely adequate to characterize risks. Such conditions include floating petroleum or other organic product on groundwater beneath occupied buildings (potential inhalation exposures), dissolved volatile organic compounds in shallow groundwater beneath occupied buildings (potential inhalation exposures), presence of vegetable gardens in contaminated soils (potential ingestion exposures) and "catch and eat" fishing activity in surface water receiving contaminated groundwater discharge. If any of these conditions or other similar conditions exist, a more detailed investigation and risk assessment may be required.

In screening environmental risks, site surface water and sediment concentrations can be compared to EPA, U. S. Corps of Engineers, and state water quality criteria for the protection of aquatic life and sediment quality criteria or guidelines. If there are no exceedences, risks to aquatic receptors are unlikely to be significant. Evaluation of potential impacts on terrestrial plants and animals at a screening level is not as simple as for the aquatic receptors. Very few, if any, benchmark soil concentrations for protection of terrestrial plants and animals have been published. Generally, if human health is protected a good case can be made that the environment will also be protected. One very significant exception to this is when a site has threatened or endangered species. In this case, benchmark concentrations become critical and must be agreed upon with the regulatory agencies. Discharges of contaminated groundwater to wetlands or the existence of threatened or endangered species on the site or in the immediate area would be indications that a screening-type environmental risk assessment may not be adequate to characterize risks. In those cases, in-depth site investigation and environmental risk assessment may be required.

A screening level risk assessment should be developed after the preliminary site characterization (including sampling and analysis) has been completed. As soon as potential receptors have been identified, it is possible to determine which pathways, if any, are appropriate for receptors at the site. An example of a complete pathway would be a buried source (disintegrating drums of contaminants), porous soil (facilitates leaching of contaminants), groundwater beneath the buried drums (method of contaminant transport), and local drinking water wells downgradient and nearby (method of human contact). If there is not a complete pathway from contaminant source to receptor, then risk for the receptor should be considered insignificant. For any complete exposure pathways, a comparison of site concentrations to risk-based benchmark concentrations, standards, or guidelines should be made and interpreted based on the scientific basis and how closely site conditions coincide with that scientific basis.

The identification of appropriate benchmarks and the interpretation of the comparison of site concentrations to these benchmarks should be conducted by trained and experienced risk assessment and toxicology professionals who are well acquainted with the regulatory standard-setting process. Experience and training in environmental health, ecological activities, or risk assessment are generally prerequisites for personnel conducting technically sound screening risk assessments.

While screening level risk assessments can be very useful in determining if risks at a particular site could be significant, they also have limitations. As with any risk assessment, the conclusions are only as good as the site characterization information. If this information is not representative of site conditions, then the risk assessment will reflect that uncertainty. In addition, the screening level assessment will generally not include site-specific exposure information (such as frequency and duration of exposures) and, therefore, should be considered an approximation of site risks. Screening level risk assessments tend to be very conservative provided all realistic exposure pathways have been addressed. Therefore, these assessments have greater utility in showing that risks are clearly insignificant or clearly significant than in determining the actual magnitude of risk or site-specific remedial requirements. Additionally, there are other methods available for evaluating sites, such as the Agency for Toxic Substances and Disease Registry (ATSDR) public health assessment process. In cases where there are unique circumstances, these alternative methods may be appropriate.

3.3.6 Presentation and Reporting

For any Phase II investigation, a report of findings will have to be prepared. The report is intended to provide the client or other readers with the salient facts of the study.

- Why the study was conducted.
- Who tasked the study.
- An outline of objectives and goals in conducting the study.
- A summary of techniques used in the investigation.
- Site information and how obtained.

- Results and interpretation of data collected.
- Summary of findings, recommendations, and/or opinions.
- Limitations of the study and techniques used; the extent to which readers can draw their own conclusions; and the extent to which readers can use the information.

How the report is prepared and how the data and findings are presented can make a tremendous difference in how the report is viewed by others and how findings may be interpreted. Numerous law suits have resulted from poorly prepared reports, and numerous endeavors undertaken because of improper statements of findings and recommendations.

More than anything else, the report should be clear, concise, and factual. As the state of the art for site assessments has evolved over the last 15 years, clients and attorneys have become ever more cognizant of the potential for disaster if the report is prepared incorrectly, or if it does not state the results in a specific way or is not worded in a particular style. This has led many clients and attorneys to work closely with the environmental specialist and assist in the report preparation, with clients and attorneys sometimes asking that the report be worded a certain way or that certain facts, findings, opinions, and recommendations be stated or not stated, etc. There is a fine line between when someone is changing the meaning of items, and when someone is asking only that certain items be worded a certain way. For example, it is common for a client to ask that recommendations not be made within a report, rather that they be presented orally or under separate cover. If the study is being conducted under regulatory scrutiny or will be submitted to a regulatory agency it is important that the report covers all of the features that the agency requires. Any omission not only will jeopardize the outcome for the client, but may also jeopardize the reputation of the person and/or firm preparing the report.

3.3.6.1 Report Outline/Format
A report needs to be prepared for each Phase II investigation. The report summarizes the results of the Phase II investigation, provides a summary of relevant findings, and possibly provides recommendations. If a Phase I Study was conducted, the information and results of such a study should be incorporated in the Phase II report.

If the Phase I was conducted by another firm or organization, these results should be referenced giving proper credit. Obviously, each report should be tailored to meet the needs of that specific situation.

Report Outline/Format While there are no national regulatory standards as to the format for a Phase II report, the environmental consulting community and the regulatory agencies have over the past decade prepared and commented on thousands of reports. From this effort the following format outline for a Phase II Report has emerged and represents a minimally acceptable outline. Table 3–10 provides an example of a Table of Contents for a report prepared based on this outline.

- Introduction
- Site background
- Scope of field activities
- Findings, opinions, and recommendations

1. Introduction This section should state the purpose and objectives for the Phase II investigation; who commissioned the study (identified as confidential if requested by the client); the reasons for the study; and any other statements regarding limitations, report setup, or format, etc., that are necessary to inform and set the backdrop for the rest of the report.

2. Site Background This section should detail information of the property or site under investigation. The level of detail provided here will depend on whether a prior study was conducted. At a minimum, the following subheadings should be included.

- Site ownership & location.
- Site description (including physical and environmental settings).
- Potential sensitive receptors.
- Site history and land use (including waste management generation and handling/disposal practices).
- Current and former uses of surrounding properties.

It is likely that an issue with the site's history, physical condition, waste handling practices, or a potential sensitive receptor created the need for a Phase II investigation to be conducted. Therefore, in preparing this section it is important that those features or issues that factored into the decision for a Phase II are presented and addressed in a manner that supports the conduct of the Phase II investigation.

In many Phase I reports, a detailed summary of information collected from federal, state, and/or local agency files or databases is given. That information is provided within a Phase II to a limited extent—mainly what has been reported before, how long ago was it reported, and who reported it are required. A cost-effective method of doing this is to provide that previous report in part or whole within an appendix. In that manner, the Phase II report text may summarize salient portions, referring to the appendix for more information.

3. Scope of Field Activities This section of the report details the scope of any exploratory, sampling and/or analysis conducted for the Phase II. Since most Phase II's are conducted to gather more data at a site to confirm or deny the presence or absence of contaminants, this section is very crucial.

Special procedures of how work was conducted, where and when it was performed, by whom, how the sampling and analysis program was designed, and why it was designed the way it was, all need to be addressed. Many investigations will have a detailed Field Program Work

TABLE 3–10   Sample Phase II Site Assessment Report Table of Contents

1.0 Introduction
2.0 Site Background
     2.1 Site Ownership and Location
     2.2 Site Description
          2.2.1 Physical Site Setting
          2.2.2 Environmental Site Setting
     2.3 Potential Sensitive Receptors
     2.4 Site History and Use
     2.5 Current and Former Uses of Surrounding Properties
3.0 Background Information Research
     3.1 Federal Information
     3.2 State Information
     3.3 Local Information
     3.4 Site Specific Information or Previous Studies
4.0 Scope of Exploration and Sampling Program
     4.1 Data Quality Objectives
     4.2 Geophysical Investigations
     4.3 Test Pit Excavations
     4.4 Soil Borings
     4.5 Monitoring Well Installations
     4.6 Site Sampling
     4.7 Site Survey
5.0 Results of Exploration and Sampling Program
     5.1 Site Hydrogeology
     5.2 Geophysical Results
     5.3 Surficial Sampling Results
     5.4 Subsurface Soil Sampling Results
     5.5 Groundwater Sampling Results
     5.6 Surface Water and/or Sediment Sampling Results
6.0 Limitations, Findings, Opinions, and Recommendations
7.0 References

List of Tables

• Adjacent Property Owners
• Summary Analytical Data Tables
• Groundwater Elevation Measurements
• Summary Table of Applicable Regulations

List of Figures

• Site Location Map (USGS Topo)
• Site Layout or Site Sketch Map
• Location of Adjacent Listed Properties
• Sampling Location Plan
• Groundwater Flow Map
• Geologic Cross-section
• Contamination Distribution
• Well Construction Schematic

Appendices

• Complete Analytical Data Packages
• Boring and Well Construction Logs

Plan, Sampling Plan, or Field Scope of Work or similar plan prepared prior to the actual task which may be referred to in the report and contained within an appendix.

4. *Results of the Field Activities*   The factual results of the investigation are presented within this section. The intent is not to derive conclusions or findings here, but simply state the facts; for example, that certain chemical compounds were detected, at what concentration within what medium, and where on the site. What the data means and how contaminants relate to site environmental or hydrogeological conditions should be presented within the findings, or opinions section (described below).

5. *Findings, Opinions, and Recommendations*   This section presents a summary of the major findings of the Phase II investigation, a listing of opinions as to the issues of concern, determination if contaminants of concern are present or not, and whether they pose a risk. The word "conclusion" should be avoided in these reports since "findings" and "opinions" have more flexibility, and do not give the perception of finality that "conclusion" carries.

Findings, opinions, and recommendations are the heart of the report and present the response to the goals of the study. For this reason, a more detailed discussion of these is contained in a later section of this manual.

3.3.6.2 *Figures and Tables*

Within any report, the use of figures and tables greatly enhances the presentation of data and ideas. The old phrase, "a picture is worth a thousand words," is never more true. Photographs of the site, sketches of the property layout, sketches of salient features central to the study, well-logs, groundwater maps, and the like are useful in providing the client and other readers with a visual understanding of the site and the findings of the study.

At the start of the report preparation, a list should be made of all the anticipated figures and tables to be developed. Preparing these early will often lead to a more efficiently prepared report (and therefore less costly), and quite often the figures and tables will make it easier to draft and formulate findings. A checklist of common illustrations and tables is provided in Table 3–10.

Figures   All figures presented within a report should contain sufficient information to stand alone; that is, the reader need not refer to the text of the report to understand what is depicted. This is even more critical if the subject figure is removed or copied by other parties. The figure should remain identifiable as to its source and should be complete enough to reduce the chance of misinterpretation.

Each figure should contain the following information; an example is provided as Figure 3–3:

- title of the figure,
- figure number,
- site name and project name,

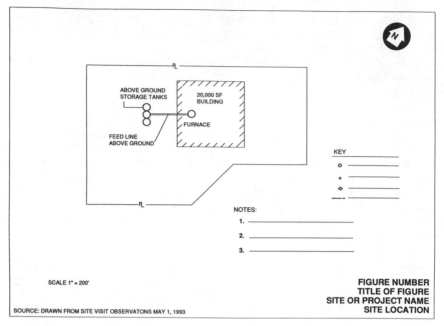

Figure 3.3

- site location,
- north arrow,
- scale bar or inch equivalent; if there is no scale or if the figure is drawn to approximate scale, then indicate that instead,
- a "Key" to the figure's use of various symbols,
- a "Notes" section, if necessary,
- a statement as to the source of the figure,
- firm's organization or name,
- the project number or other alphanumeric coding that will enable matching of the figure with the correct report or project, and
- Professional Engineer's seal, if required.

Many other items can be added to each figure, but caution must be applied to ensure that the figure does not become overcrowded with information. If the figure cannot convey the desired information to the reader in a short time, then the figure has failed in its purpose.

Tables Tables are equally important in a report as they convey a wealth of information in a small space. This is particularly important for analytical data and other numerically based information.

Like figures, tables must be properly arranged and presented if they are to meet their intended purpose. Listed below are some critical factors for tables, an example is presented as Table 3–11:

TABLE 3–11 Sample Summary of Groundwater Analytical Results
Anywhere, USA Sampling Date:

Sample Location Sample Number Lab ID # Remarks		MW02 LXX-201 XX26 —	MW02 LX-202 XX27 Duplicate	Blank LX-203 XX28 —
Volatie Organic Compounds Method 601/602 (ppb)	Method or Instrument Detection Limit (ppb)			
Acetone	5	ND	ND	ND
Chloroform	5	ND	ND	ND
Benzene	2.5	210	205	ND
Xylene	2.5	50	60	ND
Toluene	2.5	300	270	ND
Inorganic Compounds (ppm)				
Arsenic	1	2	5	ND
Barium	1	NS	NS	NS
Cadmium	1	5	8	ND

Notes:
ND Compounds not detected above the instrument detection limit
NS Not sampled for parameter indicated

- Table number,
- Table title,
- Site/project name and location,
- Date(s) of sample collection (if analytical data is being presented),
- Page number for the Table—for multiple page Tables, and
- A "Notes" section if necessary.

Since the majority of tables used for Phase II reports are analytical in nature, a discussion of them is provided below.

Table Heading/Titles: All tables must be titled and presented with sufficient detail to allow a quick understanding of the table's contents. Again, like figures, tables are commonly removed from a report and the identity of what site, media sampled, etc., may become lost if not presented on the table itself. There is no specific rule for analytical tables, but a common approach is to place the heading/title at the top of the table in a centered position. Each line could be as follows.

1st–Table Number
2nd–Site or Project Name
3rd–Table Title

Date of sample collection, method of analysis and units of measurement may be placed as part of the heading if they are consistent for the whole table or they may be included in the body of the table if they vary for different data being presented.

Table layout: The table can be either in the vertical (portrait) or horizontal (landscape) position, whichever format the data fits within. The table should be arranged with the compounds or parameters tested for down the left side (or Y-axis), and the sample points along the top (or X-axis), with the analytical results within the cells created by the matrix.

Depending on the amount of data to be tabulated or shown within a table, the table may display data for a suite of compounds (e.g., volatile organic compounds by Method 624), or summarize data for a wide variety of compounds (e.g., metals, volatiles, semi-volatile organic, etc.). Table 3–11 is an example of the latter type of data presentation.

Detection Limits: Detection limits for each compound should be given, and can be placed as either the first (left-most) data column or the last column. This is done since detection limits may vary by compound even with the same analysis.

Data to Display: Determining which data to display in tabular format has always been a problem for investigators. Presenting only those compounds where there was at least one positive hit versus all the compounds tested for can sometimes lead the reader to think that you didn't test for all compounds. Presenting all the data can, unfortunately, lead to large tables. This is generally a decision that should be made early in the study and perhaps with the client. Another question to ask is if the agency that will review the report has any requirements or preferences for data presentation. A general approach/suggestion is that the larger the project, the more data to work with, thus, the tables should be kept simple by opting for a "hits only" display.

Notes: Notes are commonly added, if not required, for most analytical data tables. The notes, usually placed at the bottom of the table, inform the reader as to the various symbols used in the table. The notes can incorporate some or all of the following items.

J   = quantitation is approximate due to limitations identified in the laboratory report (refer to Appendix if the laboratory report is attached)

ns   = Not sampled for parameter indicated

nd or * = Compounds/elements were not detected above the instrument's detection limit

3.3.6.3 Findings and Opinions

In any report of findings and opinions the question of conclusions needs to be addressed. It is generally considered that conclusions have a more significant finality to them than findings or opinions; that conclusions are themselves facts and cannot be denied or dismissed with new evidence or data. From a legal perspective, the word conclusion should be avoided in reports, as findings and opinions have more flexibility.

"Findings" are different from "Opinions." An "opinion" is a statement made based on the facts gathered from the investigation and "findings" are derived from those facts. Additionally, opinions often draw upon information that isn't necessarily connected, such as analytical data from groundwater samples and soil geology characteristics that may indicate possible pathways of migration. That the soil is porous enough to transmit a certain volume of water in a particular direction is a finding, much like the analytical data that shows a certain highly soluble compound is present in groundwater. The statement "the compound can then migrate through the soil in a certain direction and will remain above a certain concentration by the time it reaches a certain off-site sensitive receptor," becomes an opinion.

Each opinion should be supported by both factual evidence gathered during the study (and presented in the Results section) and the findings from those facts. It cannot be assumed that the reader can make the leap from facts to opinions; state all facts and findings for each opinion.

The final opinion regarding any site will likely relate to whether or not contaminants are present, what degree of risk is present and whether notification to the regulatory agencies required, and if significant, what should be done next. Ultimately, the client is asking if there is an environmental liability associated with the site. Given the likely scope of the Phase II Site Assessment, the study may be able to rule out certain issues or liabilities; at the same time there may still be data gaps, or the data may be inconclusive.

Given the diversity of sites investigated, the varying state and federal regulations, and the needs of the client requesting the Phase II Site Assessment, the number of possible opinions that could be made are endless; nevertheless, five basic levels of opinions have been identified, which are as follows.

1. No regulated compounds/chemicals were detected or noted at any concentration. Or, no evidence of a release was detected or noted.
2. No regulated compounds/chemicals were detected above the natural background levels. This would apply to inorganic species only.
3. Regulated compounds/chemicals were detected at the site, but at concentrations below standards, regulatory agency criteria, or risk-based levels and, thus, pose no risk.
4. Regulated compound/chemicals were detected at the site at concentrations above standards, regulatory agency criteria, or risk-based levels; however, these should not impact human health or the environment due to lack of sensitive receptors or exposure pathways.
5. Regulated compounds/chemicals were detected at the site at concentrations above standards, regulatory agency criteria, or risk based levels and pose a potential risk to human health or the environment.

The following are suggestions on how an opinion for each of these levels could be worded.

- Levels #1 & #2; "Clean & Green Sites"—for those sites where the opinion is that there is little or no likelihood or evidence of a problem (i.e., a release to the environment of any hazardous materials/wastes/by-products as defined by the regulations relevant to the study).

  Based on the background research and additional investigative work (e.g., subsurface and/or sampling), summarized within this report, (environmental specialists name or firm) did not find any evidence of releases at the site of hazardous materials (other wording may be better suited to the specific state or regulation), as defined by (the statute or law under which the study was conducted.)

  An alternative to "releases" would be: ". . . did not find any evidence of the presence of compounds classified as hazardous materials, as defined by . . . ." Careful wording is important here, as some compounds, such as lead, may be classified as hazardous even though they are present naturally and at concentrations that pose minimal risk.

The environmental specialist may also want to strengthen the opinion (if necessary) with the following.

  Furthermore, the current and former uses of the site are not likely to have contributed to or to have created a release of any hazardous materials to the environment.

- Levels #4 & #5; "Not Clean and Not Green"—in these situations the investigation clearly identified or detected the presence of compound/chemicals at concentrations above acceptable levels. The opinion could be:

  This investigation detected compounds or chemicals (specify the exact compounds or, if many, specify the groups) at concentrations both above background (if inorganic) and acceptable levels as set by (regulating agency or regulation). Based on our experience, it is our opinion that this site represents a potential environmental liability in that the regulating agency may dictate further actions or clean-up.

If it is felt that there is little risk due to a lack of exposure pathways or for other factors, then the following can be stated.

  However, due to the lack of sensitive receptors or exposure pathways (environmental specialists name or firm) is of the opinion that there is little likelihood for risk, based on the findings and facts presented in this report.

Obviously, all of the above-mentioned opinions need to be further supported by the facts and findings of the study and these somewhat simplistic statements will not suffice by themselves. They would all need to be adapted to each site and situation. But they should provide a starting point for developing specific final opinions.

3.3.6.4 *Making Recommendations* Recommendations within a Phase II Site Assessment report are almost as ubiquitous as some heavy metals in the environment. Clients have often been concerned that many recommendations are a ploy by consultants to create more work for themselves (i.e., keep studying the problem). Thus, it is very important to discuss "recommendations" with the client before the work begins. Do they want "recommendations" and, if yes, should they be made within the report or under separate cover?

When making any recommendation, either in the report or within a separate letter to the client, it is important to present a reasonable argument for the recommended action. Determining if any form of notification to either a federal, state, or local agency is required will be one of the more important recommendations. Clearly, notification raises the potential for liability for a private sector client.

Most other recommendations will fall into the "additional study" category. There is the fine line between the environmental professional wanting more data before a final determination, and enough data to make a reasonable opinion. If required, the additional work is likely to supplement data already collected, generally for the purpose of clarifying existing data (i.e., is there really a release, are the contaminants widespread, or are they isolated to a small area). All of the recommendations should be fully justified as to how they will add value to the clients' evaluation of the site, or if they are required by a certain regulation.

3.3.6.5 *Study and Report Limitations* Presenting a statement of limitations within the body of the Phase II Site Assessment report is done to provide the environmental specialist and others with limitations regarding interpretations that other parties can or cannot make from the facts presented.

It is common for many investigators to limit the use of the report and its findings to the client or the group paying for the study and the report. Limiting the report to just the client may provide problems for the client if the report is to be used by a lender or other party in a transaction for which the report was required. The investigator should discuss the nature of the limitations at the start of the project.

Limitations can be structured into the three following sections.

- The first section describes the name of the study, the property or site, and the client. In addition, a further statement should be included that the evaluations, assessments, and opinions contained in the report represent the environmental specialists' professional judgement/opinion as based on generally accepted engineering, scientific, and investigative practices for Phase II Site Assessments in existence at the time of the study.

This final assessment of "accepted practices" in existence at the time of the study will then require that, at least in the project file, standards or criteria or operating procedure (firm's, state or federal agency's, etc.) be identified.

- The second section would describe that the assessments and findings or opinions presented in the report are based on the activities of the environmental specialist. It should state, in a generic fashion, those activities, such as review of available background information, observations made during site visits, and the data collected from the field and analytical program as described within the report. Some investigators have further stated that the analytical data was for the detection of specific compounds above the method and instrument detection limits for only the analytical methods used.

The final section details the restrictions of use and limitations of interpretation. Examples are as follows.

"Because a limited number of samples were collected from the subject site for laboratory analysis, this report does not reflect undetected variations in chemical concentrations that may occur in the groundwater or soil and the potential presence of chemical compounds or other materials that were not analyzed for and may be present on or at the site."

"The information contained within this report and developed from the Phase II Site Assessment may not be suitable for other use without adaptation for the specific purpose intended. Any such reuse of or reliance on the information and opinions expressed in this report without adaptation shall be at the sole risk and liability of the party undertaking the reuse."

Clearly, it is possible to go overboard and give the reader the sense that the study and the report are essentially valueless. But it should also be recognized that many financial deals and property transactions potentially involving millions of dollars and people's livelihoods can be dramatically affected by the results and professional opinions of the study. In situations of perceived harm to the client (i.e., someone buys a contaminated property thought to be clean because of the report), he will likely come back to the environmental specialist to recover losses. Stigma damages must also be considered. Stigma damages are indirect or consequential damages that arise because the property is considered "tainted" by contamination. Protection in the form of limitations is of course itself limited. Contract terms and conditions and experience on similar studies (and in general) when adapted wisely and judiciously can be the most valuable protection.

REFERENCES

ATSDR Public Health Assessments, Guidance Manual, 1991

A Compendium of Superfund Field Operations Methods, USEPA, EPA/540/P/87/001, December 1987.

Environmental Data Needed for Public Health Assessments, A Guidance Manual, USDHHS, June 1994.

Groundwater Handbook, Volume II, Methodology, USEPA, EPA/625/6–90–016b, July 1991.

Guidance for Conducting Remedial Investigations and Feasibility Studies Under CERCLA, USEPA, EPA/540/G–89/004, 1988

Risk Assessment Guidance for Superfund, Volume I, Human Health Evaluation Manual, USEPA, EPA/540/1–98/002, 1989.

Subsurface Characterization and Monitoring Techniques, Volumes I and II, USEPA, EPA/625/R–93/003a&b, May 1993.

Chapter 4

PHASE III—REMEDIAL INVESTIGATION

4.1 INTRODUCTION

The preliminary site assessment and site investigation processes described so far are frequently parts of an overall procedure that, in the Superfund context, EPA calls the Remedial Investigation/Feasibility Study (RI/FS). The RI/FS is a procedure mandated by the Comprehensive Environmental Response, Compensation, and Liability Act (CERCLA, also known as Superfund) to begin the cleanup of contaminated sites. As the name implies, the purpose of the RI/FS is to develop the data necessary to plan the site remediation and to use that data to determine the feasibility of various remediation alternatives. As this methodology is well documented and provides an accepted procedure for site investigation and remediation, this manual identifies the EPA RI/FS process as primary guidance for Phase III investigations.

EPA has developed guidelines for the performance of the RI/FS entitled, *Guidance for Conducting Remedial Investigations and Feasibility Studies Under CERCLA*, USEPA, EPA540 G–89004. A flow chart of this procedured guideline is represented in Figure 4–1.

The preliminary assessment and the site investigation, corresponding to the Phase I and Phase II investigations, were described in Chapter 2 and Chapter 3, respectively. Development and screening of alternatives and detailed analysis of remedial alternatives are described in Chapter 5 as the Phase IV remedial planning, design, and implementation.

The Phase III, or Remedial Investigation, is the subject of this chapter. It is represented in Figure 4–1 by the arrows that show the interaction between two blocks: Site Investigation and Treatability Investigations, and Development and Screening of Remedial Alternatives. The objective of the remedial investigation is to provide the detailed, in-depth data necessary for choosing and implementing a remedial alternative. Information in this chapter emphasizes technical requirements for remedial

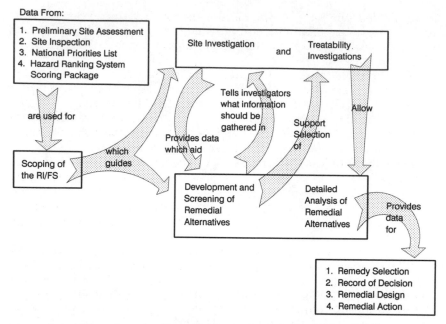

*Figure 4.1—Flow chart for Remedial Investigation/Feasibility Studies
Adapted from EPA (1988)*

investigations. In addition, community relations requirements exist which require risk communication to the public. Outstanding efforts to investigate, and design and implement remedial actions can be thwarted completely by the inability of the public to understand the health risks and risk reduction associated with the action. In fact, under Superfund requirements, community involvement through a formal community relations plan is required.

Data developed by the site and remedial investigations are used to begin screening of remedial alternatives as Phase III proceeds. For example, the alternative of bioremediation is appropriate when the contaminant is an organic compound that can be biologically converted to harmless products such as carbon dioxide or water. It may be rejected if the site investigation shows the presence of non-biodegradable or poorly biodegradable pollutants. Copper, for example, is often considered hazardous because of its total elemental concentration, and cannot be remediated by biochemical transformation. Trichloroethylene can be degraded to harmless products, but this usually occurs so slowly that bioremediation may not be practical.

The remedial investigation is often an iterative process. As possible remedial alternatives are identified, they will guide further investigation. One may propose, for example, to excavate the soils that are contaminated with copper and transfer them to a Class I landfill, while treating the remaining soils with on-site bioremediation. This could demonstrate the need for more data to precisely define the portion of the site that is copper-contaminated, so that the costs of excavation and disposal can be better estimated.

For complex sites, several iterations of investigation, sampling, and analysis may be necessary to fully define acceptable alternatives. Later iterations will include treatability investigations. If bioremediation is considered, for example, it may be necessary to test the proposed procedures in the laboratory. This will determine whether the contaminants are degradable under existing conditions, what nutrients should be added, what water content is best, and how much time is needed to reduce contaminant concentrations to non-hazardous levels. If the treatability investigation indicates that a proposed remedy is unworkable or too expensive, it will be necessary to return to the development and screening of further alternatives. These may in turn require appropriate treatability testing.

Treatment systems for excavated soil or water pumped from a contaminated aquifer often consist of a combination of unit operations. These may include chemical treatment, flocculation and sedimentation, and sand filtration, for example. Each unit operation may require testing in the laboratory to show that it will produce an effluent that is treatable by the next unit operation.

The iterations are continued until a detailed analysis of the workable remedial alternatives have been developed. The procedure should eliminate those processes which will not meet project goals, or that are obviously unacceptably expensive. The procedure results in a short list of remedies that do meet the remediation goals at reasonable cost, along with supportive data.

4.2 DATA EVALUATION

Review of the data collected is an important continuous process during the Phase III investigation. The ultimate objective of data analysis is to create a conceptually adequate and quantitatively accurate model of the site. The interim objective is to develop knowledge necessary to guide the ongoing investigation. The methods are similar to those used in data evaluation in Phase II, but are more detailed and extensive. In particular, it is possible that mathematical modeling may be employed.

The results must always be evaluated in light of the Quality Assurance/Quality Control (QA/QC) measures which were in place for the investigation program. It is tempting to accept the data at face value. But the certainty of any conclusions must be assessed only after consideration of the statistical reliability of the results and the probability of outright mistakes.

The data evaluation should include a description of the sources of contamination and ultimately create a picture of the nature and extent of the contamination. This is necessary to plan the cleanup and to determine the cost of various remedial alternatives. It will also guide the assessment of risks. Knowing the present distribution of the contaminants is also the first step in analyzing possible transport pathways and future contaminant distribution. Contaminants in surface soils may be transported on dust blown by the wind, for example, while contaminated groundwater will move as the water moves in the aquifer.

It is also necessary to know the potential receptors. If the possible impacts are limited to wildlife, for example, the study may come to different conclusions than those appropriate to a site that affects a population of children.

Ultimately, the data analysis leads to an assessment of possible effects on human health and environmental quality. Because the final object of site cleanup is mitigation of these two types of damage, it is particularly important to have them accurately defined.

4.2.1 Site Characteristics

The data provided by the Phase I, II, and III investigations must be compiled and summarized to create as complete a picture as possible, based on available data, of the physical characteristics of the site. The topography, geology, and ecology of the contaminated area will be of prime concern. The importance of specific data will vary. For example, if the site is sloped, and surface soils are contaminated with metals, it will be important to collect information on the likelihood of erosion, which could carry contaminants to other locations. It might not be important to determine subsurface and aquifer characteristics at such a site.

The physical characteristics of the site may influence implementation of some remediation alternatives. Steep or rugged terrain will limit access by heavy equipment. A high groundwater table will make excavation more expensive because the work will require pumping and treatment of contaminated water.

Knowledge of contaminant source characteristics will be of prime value in further investigation and action, and must be described in the greatest detail possible. If an underground storage tank has leaked, for example, it is necessary to know what compounds the tank contained and to determine how much product was lost, in order for the investigators to determine whether an investigation is complete. If soil analyses can show the whereabouts of 500 gallons of gasoline, for example, but 5000 gallons have been lost, it is clear that further investigation is necessary to prevent an unpleasant surprise in the cleanup process. Endless case histories can be presented of cleanup efforts which were pushed beyond deadline and beyond budget by the unanticipated discovery of forgotten buried barrels or storage tanks which were not on the property plot plans.

The nature and extent of the contamination are perhaps the most obvious site characteristics to be developed. Remedial alternatives will be constrained according to the identity of the contaminants. The costs of cleanup will rise with the amount of soil or water that is contaminated. The fate and transport of the contaminants will vary with the type of contaminant (and therefore its solubility, vapor pressure, and other physical characteristics) and the type of media (soil or water) contaminated.

Description of the nature and extent of contamination is most often in the form of maps and figures. These may be plans of the site or vertical sections of soils and aquifers. They may display isoconcentration lines describing the contaminant distribution or even the risks associated with contact with the contaminated media. Such pictorial or graphical presentations of the data are particularly desirable because they provide a powerful, concise, and easily understood summary of conditions at the site.

In the ideal case, a description of contaminant fate and transport can be developed from the description of the source characteristics and the present distribution of the contaminant. If the investigator knows when and where the release occurred, and the present location of the contaminant, then the rate, direction, and means of transport can be inferred. If contaminated groundwater is found 500 meters from a tank that leaked two years ago, then the transport rate is 250 meters per year.

Alternatively, the source characteristics can be combined with a detailed analysis of the transport phenomena to determine what transport is likely. That is, if there has been a single release from an underground tank into the groundwater, and the velocity of groundwater flow and contaminant retardation are measured, then the transport of the contaminant can be predicted.

Often the analysis of fate and transport is performed with computer models. Phenomena such as groundwater flow, vapor transport, and vadose zone migration are described by complex equations. Solving these equations for the data describing a site is commonly a lengthy iterative numerical procedure that is only practical with a computer. Many programs exist for these purposes and can be purchased "off-the-shelf." Such analyses are valuable and, in many cases, represent the only possible approach to assessing and predicting the fate of contaminants. They must, however, be used with caution. Even the best models, for reasons of economy or lack of basic knowledge, must ignore some phenomena that may contribute to transport. The investigator should be aware of all of the assumptions that were made to make the model simple enough for efficient calculation. Wherever possible, measured values for model parameters are preferred to textbook values. Often field practitioners are aware of actual data that are better than those routinely used in models.

The quality of the model predictions is critically dependent of the quality of the data. An assessment of transport by vadose zone flow, for example, may err by a large margin if an impermeable lens of clay underlies part of the site and is not detected in the geological evaluation.

Models are best used in an iterative manner. Collecting the data and tabulating it in the form required for input is often an excellent exercise for making the investigator familiar with its completeness. This alone may show what data should be collected during the next visit to the site. The models may also be used for sensitivity analysis. For example, if an input parameter is known to within a given range of values, then the model may be run with inputs throughout this range. If the result changes only slightly for this range of values, then the investigator can conclude that that parameter is sufficiently well known. However, if the uncertainty in the input parameter generates large uncertainties in the results, it will be necessary to go back to the field to get better data.

4.3 ASSESSMENT OF RISKS

Site investigation must produce a description of the hazards the contamination poses to public health and natural ecosystems, and must predict how those hazards will be mitigated by each of the proposed remedial alternatives. Alternatives that do not reduce risks to acceptable levels will be rejected or modified.

4.3.1 Identification of Contaminants

The type and amount of contaminant present must be known. When the compounds have been identified, a library search can provide data on their intrinsic toxicity and the health risks. Compounds with high vapor pressures, for example, are more dangerous because people on or near the site may breath the vapors. Soluble materials are more likely to enter the groundwater and leave the site, while those which adsorb tightly to soils will remain in place.

Identification may not be an easy task. Where the contaminant consists of one or a few compounds, instrumental methods may provide unambiguous data on how much is present. Many toxic contaminants, however, consist of complex mixtures. If crude oil is spilled, for example, thousands of different compounds may be present. Each will have its own physical and toxicological characteristics. Each will be degraded by soil microorganisms at a different rate. Many may be difficult to analyze against a sample matrix containing many other compounds that can interfere. Some, like the polycyclic aromatic hydrocarbons, may be of concern even near the limits of detection, where the data are unreliable.

4.3.2 Exposure Assessment

Assessment of probable human exposures is key to the choice of a remediation method. Information on the characteristics of the sources allows identification of which media will be contaminated. Transport processes can be described for each media. Contaminated soil, for example, may be transported by wind. Contaminated groundwater may be transported by flow through aquifers, or may reach the surface at rivers, streams, or seeps.

When the points of possible human contact are identified, each possible exposure route must be evaluated. Vapors and dust can be inhaled, so respiratory exposure is of concern. Contaminated dust can also be ingested by people eating with dirty hands or by children tasting soils.

Sources, media, transport pathways, and exposure routes combine to produce the means by which individuals may be injured by contaminants. A necessary and perhaps most difficult step is to quantify each of these. Together, they determine the likely dose for exposed individuals.

4.3.3 Toxicity Assessment

The best estimates of likely dose to people can be combined with knowledge of toxicity to produce a toxicity assessment. The literature provides a description of the type of health damage associated with the contaminant. It is necessary to also describe the variation of health effects with dose. Every contaminant will have a minimum dose below which its health effects are negligible, or at least acceptable to regulators. Above this level, effects may gradually become worse. Combined with the exposure assessment, knowledge of the response versus dose relationship will allow prediction of health effects.

Information on the health effects of toxic substances is often incomplete and controversial. The toxicity assessment should include an evaluation of the reliability of the health information.

4.3.4 Risk Characterization

A summary and critical evaluation of contaminant identification and exposure and toxicity assessment constitutes the risk characterization for the site. Such summaries should be prepared for each of the remedial alternatives being considered, so that the health benefits of each can be compared. The "no action" alternative must be included as the baseline against which all other alternatives are compared.

Assessment of risks should be carried on throughout the remedial investigation phase. It is an important element of the iterative process by which a complete database for the site is compiled. Evaluation of transport phenomena, for example, may show that information on processes at the site are incomplete, guiding further data collection. If it is found that contaminants are entering a stream, the investigator must ask how many people live near the length of the stream in which concentrations will be significant.

The demographics of the affected population will also be important. If the affected area is residential, or includes a school site, children will be among the population of concern. They are more susceptible to toxins, and may be affected by compounds that can interfere with growth and development. Alternatively, the affected population might be a military base, with only an adult, transient population, which would be more resistant and experience shorter periods of exposure.

4.4 TREATABILITY STUDIES

Treatability studies are part of the iterative process that moves back and forth between site investigation and remedial alternative development. Site investigation provides an initial description of the problem and a remedial alternative is suggested. But the effectiveness of that alternative is often uncertain. Chemical and/or biological treatments of soils, for example, are greatly affected by the composition of the soils. These effects are difficult to predict and can only be evaluated by testing. A treatability test employs the proposed remedy on a sample of site soils. Bench scale testing is done in the laboratory and pilot scale testing is done on a larger scale, usually at the site.

The treatability tests determine whether the proposed technology will work. They also determine parameters that are important to remedial costs, such as required chemical dosages or the necessary time. In many cases, the data also constitute an optimization study. Optimization is necessary where the most appropriate values of parameters controlling remediation are not known. Bioremediation of soils often requires the addition of water and nutrients, for example, but the amount required will vary from soil to soil. The treatability test should be used to determine the best water content for two reasons. First, if the process is not optimized when the test is done, the alternative may be rejected even though the optimal version of the process is the best remedial option. Second, optimal procedures must be determined so that the remediation can be performed with best efficiency.

Two kinds of data determine whether treatability testing should be done. Data describing the treatment process are available from the literature or from vendors. They should include reports on applicability, past experience, and dependability. The second set of data is that from the site investigation. If the investigator knows the situations in which the treatment is applicable, and can clearly determine that the site in question is one of these, treatability testing will not be necessary.

4.4.1 Bench- and Pilot-Scale Testing

Bench-scale testing is conducted with small amounts of material in a laboratory. The flasks or test tubes used as reactors are cheap and it is relatively easy to do large numbers of tests. This allows extensive optimization studies because tests can be done for many values of a parameter.

Bench-scale testing also allows precise control of variables that influence test results. Temperature, for example, may be important, and it can be set to desired values in incubators holding the test samples.

Pilot-scale testing is an attempt to mimic the performance of the remedial alternative to the greatest degree practical. The pilot system is made as small as possible, to minimize costs, but is large enough so that deviations in performance from the full-size system are not expected. Pilot-scale testing may be done in a laboratory setting but is more commonly performed on-site.

Pilot-scale testing has the disadvantages of being more expensive and less flexible. Because the units are far larger than those used in bench-scale testing and must have the characteristics of the full-scale system, they are much more expensive. Design and construction of the pilot system may be time consuming. Almost always, only a single unit is built. This means that side-by-side comparisons of the effects of certain parameters are not possible.

The large size and possible on-site location of the pilot-scale system means that it is more difficult to control operating parameters. If the investigator wishes to compare its effectiveness on two soil samples, for example, the tests must be done sequentially. But if a weather change causes the second test to be done at a lower temperature, the investigator may be left wondering whether the difference in performance is due to the difference in soils or due to the temperature change. Evaluating the causes of system variability may be difficult.

However, the changing conditions encountered by the pilot-scale system will be those likely for the full-scale system. The tests may, therefore, be more realistic even as they are less easily analyzed.

4.4.2 Work Plan Preparation

The elements of work plans that should be written before treatability testing have been summarized by EPA (Tables 4–1 and 4–2). Many of the requirements are those appropriate to maintaining quality in experimental work: clear statement of objectives and methods, description of the equipment and procedures used, quality assurance for data and data management, and personnel health and safety.

TABLE 4–1   Suggested Format for Bench-Scale Work Plan

1. Project description and site background
2. Remediation technology description
3. Test objectives
4. Specialized equipment and materials
5. Laboratory test procedures
6. Treatability test plan matrix and parameters to measure
7. Analytical methods
8. Data management
9. Data analysis and interpretation
10. Health and safety
11. Residuals management

TABLE 4–2   Suggested Format for Pilot-Scale Work Plan

 1. Project description and site background
 2. Remediation technology description
 3. Test objectives
 4. Pilot plant installation and startup
 5. Pilot plant operation and maintenance procedures
 6. Parameters to be tested
 7. Sampling plan
 8. Analytical methods
 9. Data management
10. Data analysis and interpretation
11. Health and safety
12. Residuals management

Both kinds of testing will require plans for residuals management. The materials being tested are by definition hazardous, and proper disposal techniques must be utilized. Pilot-scale testing also requires plans for building or installing the equipment, which may be large and elaborate.

4.4.3 Report

The results of treatability testing should provide an evaluation of the candidate technology in terms of effectiveness, implementability, and cost. Operating procedures and conditions should be carefully documented so that they will provide useful guidance to implementation of full-scale procedures. If possible, a mathematical model of the process accounting for the effects of scale should be developed.

4.5 REMEDIAL INVESTIGATION REPORT

Phase III, or remedial investigation reports, should be similar to those for Phase II. Indeed, they will contain much of the same information, with the new material developed in Phase III added. A new evaluation of overall results, including the insights developed in Phase III, will be included.

Site characterization will be more complete. In particular, it will include the data necessary to evaluate and compare remedial alternatives. Data presentation may be similar to that in the Phase II report and many of the tables, figures, and maps will be repeated. Either conceptual or mathematical models will be included, leading from site characterization through fate and transport analysis to exposure assessment for each of the contemplated remedial alternatives.

Chapter 5

PHASE IV—REMEDIAL PLANNING, DESIGN, AND IMPLEMENTATION

5.1 INTRODUCTION

At the conclusion of the remedial investigation, a feasibility study is undertaken. That study should provide the recommendations for the most cost effective remedial actions to protect the public health and environment from the hazardous wastes present at the site. Using the site and remedial investigation results and the risk assessment information, the feasibility study can be conducted. It includes remedial planning, design, and implementation.

Feasibility studies are generally prepared to meet criteria established by EPA for activities carried out under CERCLA. Information contained in this chapter gives a broad overview of these criteria as they pertain to feasibility studies. For sites that do not fall under CERCLA, these criteria are not strictly applicable; but it is still advisable to meet the substance of the requirements if regulatory approval of the remedial activities is to be obtained or if cost recovery litigation is anticipated.

Remedial planning includes thorough development of remedial action alternatives, screening of alternatives, alternative evaluation, and preparation of a feasibility study report. A pre-design report is prepared based on the remedial planning results. When the remedial design has been approved by the appropriate agencies, implementation of the design can proceed.

5.2 ESTABLISHMENT OF REMEDIAL ACTION OBJECTIVES AND CRITERIA

Development and selection of the site remedial action is dependent on the selected objectives of protection of public health, surface water quality, groundwater quality, and air quality. When these objectives are determined, candidate remedial action alternatives can be developed and measured against criteria such as reliability and effectiveness, implementability, operation and maintenance requirements, and costs.

The first step is to establish remedial action goals for each objective based on reducing the level of endangerment to acceptable levels. This process will result in the selection of specific objectives for surface water quality, groundwater quality, and air quality. The process of translating objectives into definitive criteria is a well-established process in engineering design and is typified by performance specifications that, when achieved, assure that the objectives are met.

5.3 DEVELOPMENT OF REMEDIAL ACTION ALTERNATIVES

The remedial action alternatives developed must be appropriate to the site for each pathway and contaminated environmental medium. The identified remedial alternatives also must be consistent with the objectives and criteria developed above. They also must be directly related to the quantitative and qualitative data on contaminants, their location, behavior, and fate under site-specific conditions (soils, geology, etc.), as determined previously.

At any site, the generic remedial action alternatives can be categorized as follows:

- no action,
- containment,
- on-site/in situ remedial treatment, and
- off-site treatment and disposal.

The final list of generic remedial alternatives can then be expanded from general concepts to specific remedial action alternatives by identifying applicable alternative technologies and assembling these technologies into various combinations of operable units to form remedial alternatives for comparative screening.

5.4 SCREENING OF ALTERNATIVES

The final list of viable remedial action alternatives can be reduced, through prescreening procedures, to the "most practicable" alternatives in terms of cost, effectiveness, and technical feasibility.

5.4.1 Cost

"Conceptual" estimates, having an accuracy in the range of plus or minus 50%, are normally used to roughly determine economic feasibility.

Capital and operations and maintenance costs should both be estimated, and distributed over time for comparisons. Present value should be calculated based on historical trends of interest rates.

5.4.2 Effectiveness

Estimates of both long and short term environmental and public health impacts should include: environmental and public health impacts during implementation, degree of achievement of regulatory clean-up goals, and mitigation of the threat of harm to public health, welfare, or the environment.

5.4.3 Technical Feasibility

Each alternative must be considered for feasibility and reliability. Technical feasibility refers to the ability to construct, reliably operate, and meet technology-specific regulations.

5.5 ALTERNATIVE EVALUATION

The alternative remedies that pass the initial screening process should then be evaluated in detail. Nine evaluation criteria are considered:

- overall protection of human health and the environment,
- compliance with relevant regulations (ARARs),
- long term effectiveness,
- reduction of toxicity, mobility, and volume,
- short term effectiveness,
- implementability,
- cost,
- state acceptance, and
- community acceptance.

A brief description of the technical components of each of these criteria is contained in Figure 5-1. State acceptance and community acceptance are not strict technical criteria, but are based on the ability of the professional preparing the feasibility study to communicate the technical appropriateness and community benefits of the alternatives.

The following discussion points out the main aspects of criteria one through seven above that are used in the technical evaluation of alternatives.

5.5.1 Overall Protection of Human Health and the Environment

This criterion provides an overall assessment of whether each alternative adequately protects human health and the environment. The overall assessment focuses on whether an alternative would achieve adequate

OVERALL PROTECTION OF HUMAN HEALTH AND THE ENVIRONMENT

How alternative provides human health and environment protection

COMPLIANCE WITH ARARs

Compliance with chemical-specific ARARs

Compliance with action-specific ARARs

Compliance with location-specific ARARs

LONG TERM EFFECTIVENESS

Magnitude of residual risk

Adequacy and reliability of controls

REDUCTION OF TOXICITY, MOBILITY, AND VOLUME

Treatment process used and materials treated

Amount of hazardous materials destroyed or treated

Degree of expected reduction in toxicity, mobility, and volume

Degree to which treatment is irreversible

Type and quantity of residuals remaining after treatment

SHORT TERM EFFECTIVENESS

Protection of community during remedial actions

Protection of workers during remedial actions

Environmental impacts

Time until remedial action objectives are achieved

IMPLEMENTABILITY

Ability to construct and operate the technology

Reliability of the technology

Ease of undertaking additional remedial actions, if necessary

Ability to obtain approvals from agencies

Coordination with agencies

Availability of off-site treatment, storage, and disposal services

Availability of necessary equipment and specialists

Availability of prospective technologies

COST

Capital costs

Operating & Maintenance

Present worth

STATE ACCEPTANCE

COMMUNITY ACCEPTANCE

Figure 5.1—Alternatives Evaluating Criteria

132

protection and how site risks would be eliminated, reduced, or controlled through treatment, engineering, or institutional controls. This evaluation also allows for consideration of any unacceptable short term or cross media impacts.

5.5.2 Compliance with ARARs

This criterion is used to determine how each alternative complies with federal and state ARARS. Action-specific ARARs for each alternative are identified. Discussion should also be provided as to whether chemical-specific and location-specific ARARs are met.

5.5.3 Long Term Effectiveness

This criterion addresses the results of a remedial action in terms of the risk remaining at the site after response objectives have been met. The primary focus of this evaluation is the effectiveness of the controls that will be applied to manage the risk posed by treatment residuals or untreated wastes. The magnitude and potential of future exposure to residual contaminants, the long term reliability of continued protection, and the potential need to replace the technical components of the alternative are considered.

5.5.4 Reduction of Toxicity, Mobility, and Volume

This criterion assesses the degree to which hazardous substances would be treated to permanently and significantly reduce toxicity, mobility, or volume. Each alternative should be evaluated on its ability to reduce the principal threats at the site through the destruction of toxic contaminants, the irreversible reduction of contaminant mobility, or the reduction of the total volume of contaminated material.

5.5.5 Short Term Effectiveness

The short term effectiveness criterion addresses the effects of an alternative during the construction and implementation phases of the work until the remedial action objectives have been met. These effects include the protection of workers and the community during construction and implementation, environmental impacts that might result from construction or implementation, and the length of time until the remedial action objectives are achieved.

5.5.6 Implementability

The implementability criterion addresses the technical and administrative feasibility of implementing an alternative and the availability of required services and materials. Technical feasibility encompasses the technical difficulties and unknowns associated with the construction and operation of alternatives, the reliability of the technologies and the likelihood that technical problems will lead to schedule delays, the ease of undertaking additional remedial actions to address the other potential sources on site, and monitoring requirements. Administrative feasibility includes coordination with government offices and agencies, for example, to obtain permits. Services and materials includes the availability of offsite treatment, storage, and disposal facilities; the availability of necessary equipment and specialists; the ability to obtain competitive bids; and the availability of prospective technologies.

5.5.7 Cost

The cost evaluation of each alternative includes capital costs, annual operation and maintenance (O&M) costs, and a present worth analysis. The cost estimates for the detailed analysis should provide a range of accuracy from −30 to +50 percent, based on the available site data.

The capital costs include the direct and indirect costs associated with the construction of an alternative. Direct capital costs are expenditures for material, equipment, and labor during installation. Indirect costs are expenditures for engineering, financial, legal, and other services.

The annual O&M costs are the post-construction costs necessary to ensure the alternative's continued effectiveness, which include maintenance materials and labor costs, sampling and analysis costs, and administrative costs.

The present worth analysis is used to evaluate expenditures that occur over different time periods by discounting the amount of money that, if invested during the current year, would cover all costs associated with the remedial action over its planned life. The maximum lifetime generally considered for each alternative is 30 years.

5.6 COST EFFECTIVENESS ANALYSIS

To complete the alternative evaluation, the adequacy of the alternatives in relation to the response objectives and criteria, the effectiveness of each of the practicable remedial alternatives, as well as their composite cost effectiveness must be determined and ranked by evaluating each of the alternatives' relative ability to meet the goals of each of the seven technical criteria and the two acceptance criteria. The purpose of the comparative analysis is to identify the advantages and disadvantages of each alternative relative to one another, pointing out key trade-offs between alternatives. Cost measures (such as capital, operations and maintenance, and present value) and effectiveness measures (such as level of cleanup achieved, durability, implementability, human exposures, environmental benefits, adverse environmental impacts, and institutional issues) should be fully compared. A semi-quantitative ranking scheme can be used for the comparisons. The lowest cost alternative that is technically feasible and reliable and effectively mitigates and minimizes damages to and provides adequate protection of public health, welfare, or environment can then be recommended.

5.7 PREPARATION OF FEASIBILITY STUDY REPORT

A feasibility study report can be prepared using the above data and information. An example of the report outline is shown in Table 5–1. Each chapter should describe the work performed and detail findings, interpretations, summations, conclusions, and recommendations. Supporting data in the form of calculations, tables, figures, charts, diagrams, and drawings should be used to the extent possible.

5.8 PREPARATION OF PRE-DESIGN REPORT

After acceptance of the feasibility study's recommended alternative, a pre-design report is prepared. The pre-design report is a stand-alone document with four major sections:

- site description, characterization, and factors that affected the selection of the remedy,
- conceptual design and mandatory design criteria,
- implementation aspects, and
- appendices.

TABLE 5–1.   Feasibility Study Report Format

1.0 Introduction
    1.1 Site Background Information
    1.2 Nature and Extent of Problems
    1.3 Objectives of Remedial Action
2.0 Initial Screening of Remedial Action Technologies
    2.1 Technical Criteria
    2.2 Environmental/Public Health Criteria
    2.3 Institutional Criteria
    2.4 Other Screening Criteria
    2.5 Cost Criteria
    2.6 Development of Remedial Action Alternatives
3.0 Remedial Action Alternatives
    3.1 Alternative 1 (No Action)
    3.2 Alternative 2
    3.x Alternative x
4.0 Detailed Analysis of Remedial Action Alternatives
    4.1 Cost Analysis
    4.2 Non-cost Criteria Analysis
        4.2.1 Technical Feasibility
        4.2.2 Environmental Evaluation
        4.2.3 Institutional Requirements
        4.2.4 Public Health Analysis
    4.3 Cost-effectiveness Analysis
5.0 Recommended Remedial Action
References
Appendices

The site should be described in terms of overall features, historical events, environmental setting, potential receptors, and hazardous substances. The selected remedial action should be presented along with a summary of the key factors leading to its selection. These factors include regulatory, technical, cost, institutional, public health, and environmental issues and site-specific test results.

The conceptual design section includes a description of the remedial action, performance requirements, design criteria process flow diagrams, equipment list, site map with preliminary layouts, disposal areas, borrow areas, transportation routes, operation and maintenance requirements, monitoring requirements, cost estimates, and schedules.

Implementation aspects include special technical problems, additional engineering data that may be required, permits and regulatory requirements, access, easements, rights of way, health and safety requirements, a community relations strategy, and a preliminary work plan.

Supporting data in the form of calculations, tables, figures, charts, diagrams, and drawings should be contained in appendices.

5.9 IMPLEMENTATION PROCEDURES AND SCHEDULE

The primary objective of specifying implementation procedures and schedules is to establish the organizational structure and a time sequence of actions for remedial actions. Other elements important to implementation are management and administrative considerations.

The organization structure for remedial actions refers to the entities responsible for performing the tasks set forth in the recommended plan. A logical split of organizations is by function. Thus, both administrative and operational entities must be defined in the implementation schedule.

The implementation schedule must also contain details of expenditures, both capital and operation and maintenance costs. A time sequence within which designated entities will establish standards and other means of control must be included in the implementation schedule.

REFERENCES

Guidance for Conducting Remedial Investigations and Feasibility Studies Under CERCLA, USEPA, EPA/540/G–89/004, 1988.

INDEX

139

89; sampling plan refinement 69-70

Sanborn maps: Fisherman's Terminal area (San Francisco) 30-33; Map Legend 28-29

SARA (Superfund Amendments and Reauthorization Act): landowner liability under 4-5, 18-19. *See also* Legal Risk Management; CERCLA

Schedules, legal aspects of 12. *See also* Legal risk management

Scope of work, Phase II: final, development of 64-65; scope vs. purpose 60-61

Screening level risk assessment (Phase II) 103-105

SCS (Soil Conservation Service): aerial photographs from 25

Septic systems/drain fields: in site reconnaissance 41

Sewer lines: in site reconnaissance 40

Site history evaluation: aerial photographs 22-25; archival records/manuscripts 35-37; atlases, real estate 27; building permits/plans 34-35; city street directories 34; criteria for 21; data sources for 22; maps 25-27, 28-33; textual records 35; title searches 27, 34. *See also* Maps; Photographs, aerial

Site reconnaissance: buildings/structures 41-43; documentation, visual/photographic 37-38; documentation, written 38; hazardous substances 43-44; personal interviews 38; Phase I report 52-54; physical features/surface conditions 39-40; storage tanks (UST/AST) 45-46; surface water conditions 40; utilities 40-41; waste disposal methods 45; waste management practices 44-45. *See also* Phase II site investigation (general)

Soil gas surveys (Phase II) 74, 76

Soil investigation, invasive techniques of 78-79

Soil sampling, methods of 85

Spills/spill reports: in site reconnaissance 43, 44, 45

Stained soil: in site reconnaissance 39

State agencies (as data sources) 50. *See also* Data sources, state/local

Storage tanks (UST/AST): leakage, records of 50; in site reconnaissance 45-46

Stream hydraulics 95

Sumps: in site reconnaissance 43

Superfund. *See* SARA (Superfund Amendments and Reauthorization Act)

Surface characteristics: in site reconnaissance 39-40

Surface water conditions: in site reconnaissance 40

Surveys, metes and bounds 21

Tax records 50-51

Test pits 83, 85

Third Party Defense 18. *See also* Liability

Title searches 27, 34

Toxicity assessment (Phase III) 125

Transaction Screen Process (ASTM) 5

Treatability studies (Phase III): introduction 126; bench- and pilot-scale testing 126-127; report 128; work plan reparation/format 127-128

TRIS (Toxic Release Inventory System) 49

TSCA (Toxic Substances Control Act) 49

UCC (Uniform Commercial Code): and contractual schedules 12

Urea formaldehyde foam (UFF): in site reconnaissance 42

USGS (U.S. Geological Survey): aerial photographs from 24

Utilities: in site reconnaissance 40-41

Vegetative distress: in site reconnaissance 39, 44

Volatile organic compounds (VOCs): in Phase II sampling plan 70

Waste disposal methods: in site reconnaissance 45

Water lines: in site reconnaissance 40

Waterbodies (rivers, lakes): in site reconnaissance 40

Wells: in site reconnaissance 41

Wetlands: in site reconnaissance 40